The New Flatlanders

THE NEW FLATLANDERS

A Seeker's Guide to the Theory of Everything

ERIC MIDDLETON

TEMPLETON FOUNDATION PRESS
PHILADELPHIA

Templeton Foundation Press
300 Conshohocken State Road
Suite 670
West Conshohocken, PA 19428
www.templetonpress.org

2007 Templeton Foundation Press Edition
Originally Published by Highland Books, Surrey, UK, 2002

Templeton Foundation Press helps intellectual leaders and others learn about science research on aspects of realities, invisible and intangible. Spiritual realities include unlimited love, accelerating creativity, worship, and the benefits of purpose in persons and in the cosmos.

Designed and typeset by Gopa & Ted2, Inc.
Fractal image on back cover courtesy of Brad Johnson

Library of Congress Cataloging-in-Publication Data

Middleton, Eric.
The new flatlanders : a seeker's guide to the theory of everything / Eric Middleton.
— Templeton Foundation Press ed.
p. cm.
Originally published: Surrey, UK : Highland Books, 2002.
Includes bibliographical references and index.
ISBN 978-1-59947-123-5 (pbk. : alk. paper) 1. Religion and science. 2. Science—Miscellanea. 3. Apologetics. I. Title.
BL240.3.M53 2007
215—dc22
2007005769

Printed in the United States of America

07 08 09 10 11 12   10 9 8 7 6 5 4 3 2 1

This book is dedicated to all those who were willing to become "Flatlanders," who enriched us with their sharing and who encouraged me to write this book.

Contents

Acknowledgments

My thanks are due to all who were willing to become "Flatlanders." It was their story that they asked me to write, and that spread out in circles to many others. These successive enquirers came to unwrap the mystery in their own way. Yet the pattern proved remarkably similar for students of all ages, prisoners and titled folk, young and old. The "Flatland" name came from a nineteenth-century parable-story.

My grateful thanks must be given to Isobel Stevenson for her careful editing of the text.

I am most grateful to St. John's College, Durham University (Principal the Reverend Dr. David Wilkinson) for its continuing support and encouragement ever since my time as a Fellow of the College.

The New Flatlanders

Introduction

The view from the stone circle

Angus lent back comfortably against the warm stone. "You know, here's something really weird about this place. Once when I was up here, it started pouring with rain, but as soon as I got outside the stone circle it was quite dry. And that ley line across the field—why doesn't the snow settle there and why do the sheep line up around it the way they do?"

Melissa nodded. "I've heard that horses and dogs won't come near here, unless you force them to."

But Richard, as usual, was skeptical: "Come off it. It's just a set of markers put up around an old plague pit, or maybe even just a bunch of rocks on the side of a hill."

"It's got to be more than just rocks," said Ruth. "There is a pattern to the way they're arranged. If you look carefully, you can even see what might be some sort of avenue of stones leading to that spring down there. Probably this was some sort of shrine, where they carried out sacrifices—maybe even human ones! I am so glad that science has disproved everything to do with religion, so we can come up here without risking having our throats cut!"

"So what you're saying is that 'it's just rocks that humans messed with,'" protested Melissa. "But then why do Angus and I feel strange things here? Science doesn't have all the explanations—it can't explain how my Ouija board and Tarot cards work, but they really help me make decisions."

"Like when to put in another nose ring or change your hair color again," teased Angus.

"Not fair," interrupted Richard. "She means more than that. And there are weird things that happen. Watch this." He hauled out a length of wire that he had twisted into an L-shape. "Watch." He walked slowly across the stone circle, holding the angle of the L loosely in his hands. The ends of the wire dipped as he approached the center.

"You moved your hands," said Angus.

"No, I didn't. Try it yourself. "

"OK." Angus took the rod carefully in his hand, holding it the same way Richard had done, and set off across the circle. Suddenly the wire twisted down.

"Oh, please let me try," cried Melissa. She too felt the wires turn in her hand as she approached the center, and returned glowing to the group sitting in the shade of one of the stones. "It works," she said. "You saw it. That's an ancient power, more than science."

"Not necessarily," said Richard, folding up the rod and returning it to his pocket. "Science is also 'an ancient power,' if you think about it, and scientists can measure what we felt there. I've heard that at the Rollright Stones near Oxford they did find differences in the electromagnetic potential at some places in the circle—though they don't know why yet. And some biologists say that stags use their antlers like divining rods to detect water. I don't know the details of how it works, but I think science will be able to explain it one day."

"It's interesting," said Angus. "Mel and I are interested in all this ancient powers and magic stuff, and you and Ruth think that science debunks it all. We keep having discussions like this, but we never really settle anything. I wish we could argue these things through properly, not just touching on bits and pieces like we do now."

"I've heard there is a guy called Eric who sometimes leads a few seminars on the sort of stuff we have been talking about today. Covers everything from multiple-dimensions to Mel's 'feelings,' the paranormal, superstrings, relativity. I'll see if I can find out anything about it, and whether it's worth our doing together."

. . . . .

And that was how our group began.

Our first meeting was a brainstorming session. The students were suspicious of me, because of the large collection of books on religion (as well as science) in my study. They wanted a say in laying out the ground rules for our seminar.

"We do not want to talk about religion," Richard emphasized. "You know where you are with science. You can touch, see, measure, visualize, test, deal with facts, repeat experiments, and readily prove ideas. With religious beliefs you can never be certain."

"Yes," Ruth agreed, but Melissa and Angus seemed a little more hesitant. They weren't into religion, they explained, but they were interested in the spiritual dimension. How did that fit in with Richard's view of science?

I agreed with them that "religion" was a word to be avoided, but I suggested that one of the issues we consider was whether there might be more to science than Richard thought. After all, I pointed out, his view of science was rather dated, rooted in the nineteenth century, and he might want to update his thinking too.

Plus, I added, Richard's acceptance of science was also to some extent a statement of faith. The mathematics involved in modern physics is so amazingly complex that we have to take a lot of what scientists say on trust. They might even find some of the science so weird that they might have difficulty suspending their disbelief when discussing it. We'd probably need to consider whether we might see something similar in relation to the types of things we might call transcendent, things that go beyond the merely physical.

As the brainstorming session continued, it also became clear that the issues the students wanted to discuss were not purely intellectual ones. They had personal worries about such things as life after death—was there anything else? Did we just cease to exist? If we carried on in some other life, what difference did it make to our lives now? What about the feeling they sometimes had that some things were "meant to happen"—how was that possible if everything just happened randomly?

Gradually, the group defined its focus as looking for an understanding of physical reality that would help them cope with the fact that their experience of life involved more than just the physical. They wanted their lives to be more than random examples of biological life arising by chance on an obscure planet, but they were inclined to think this was wishful thinking—unless there was some reality to the spiritual dimension.

We duly set our course to try and find the beginnings of a working theory of "everything."

In what follows, I will not present our discussions as a dialogue with particular voices. Instead, I will take the questions that emerged from the group and use them to frame the account of what we learned as we worked together. I brought to the discussion my background in various branches of science and my faith. They brought a questioning intelligence and a willingness to at least consider the possibility that there is more to life than meets the eye.

1. The Universe and Beyond: How Did It All Begin?

There may be a good deal more to the universe than meets the eye, even the eye of faith of the cosmologist. —JOHN D. BARROW[1]

THE MORE I thought about what we were undertaking, the more ludicrous it seemed. How were we, a small group of students and a teacher, going to explore the nature of the universe and our place in it? The scope of our study was alarming. So was the fact that whatever answer we came up with would be bound to have an effect on how we saw ourselves and how we lived our lives.

It seems that the students were having similar misgivings about the scope of our project, for their first question was guaranteed to remind me of the scale of the problem.

How big is the universe?

So big that it is difficult to grasp the immense distances we need to think about. When we walk ten miles, we know what that distance feels like. Driving the same distance in a car makes it feel shorter, and if we are in a plane traveling the 5,071 miles between London and Beijing, those ten miles pass completely unnoticed. Yet even that long journey is short compared to the 300,000 miles to get us to the moon.

But the moon is actually very close compared to the sun, which is 93,000,000 miles away. There are too many zeros there for us to be comfortable with, even if we rephrase the number as ninety-three million. So we put it into a unit we can understand: it takes light just over eight minutes to travel from the sun to the earth. But the units pile on—

minutes, hours, days, years—and after a mere 4¼ light-years we reach the nearest star.

But our sun and that nearest star are only two of the many millions of suns whirling around in an immense spiral with our sun halfway along one of the rotating arms. Gazing up at the night sky on a clear night, we can see part of this huge galaxy in the form of the Milky Way, an immense swath of stars above us. Seen from above, it would look like the other spiral galaxies we can see through a telescope; from the side, it would look like two inverted dinner plates—or perhaps a fried egg—measuring 100,000 light-years across.

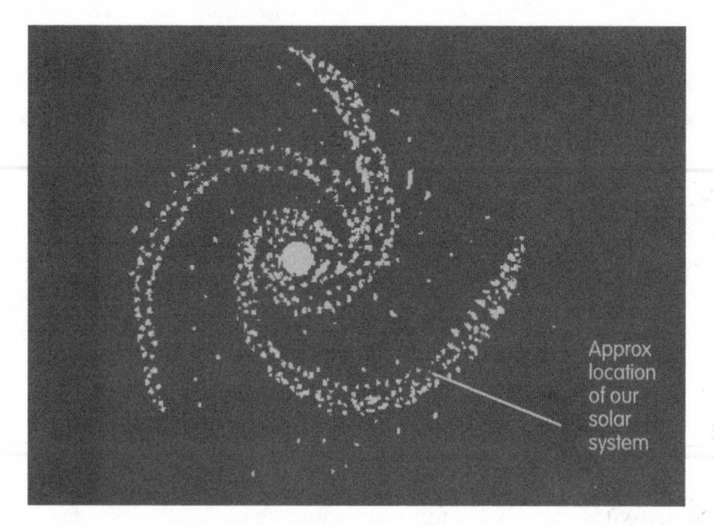

FIGURE 1.1 MILKY WAY GALAXY SEEN FROM ABOVE

You said that it looked "like other galaxies." How many galaxies are there?

There's no shortage of them! The nearest is Andromeda, over 2.3 million light-years away from our own galaxy, and that is only one of a group of twelve or so that make up our local cluster of galaxies, circling round one another in the immensity of space. But there are other clusters also, making a local supercluster—and further superclusters of galaxies as far as telescopes can record—as least as far as thirteen to fourteen thousand million light-years away. And that is our current best estimate for the size of the universe!

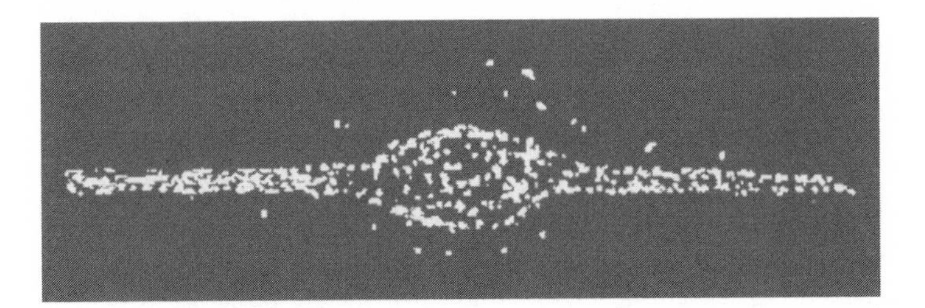

FIGURE 1.2 MILKY WAY GALAXY SEEN FROM THE SIDE

Given those distances, it doesn't sound as if we'll ever be able to reach another galaxy, does it?

Not unless we discover some remarkably different way of traveling. And the problem isn't just the distance: it's also the fact that the distance keeps growing; the furthest clusters of galaxies are moving away from us at enormous speeds.

How do we know that? In 1929 an astronomer called Edwin Hubble drew an analogy to the way in which the whistle of a train or the siren of a police car changes pitch as it moves away from us and the way the lines on the spectrum of light from distant galaxies are shifted to the red end. The further away the galaxy, the bigger the red shift. He proved that the galaxies are moving away from us, and that we actually live in an expanding universe.

To try and visualize this, imagine that the three dimensions of space are represented by the two-dimensional surface of a balloon with dots

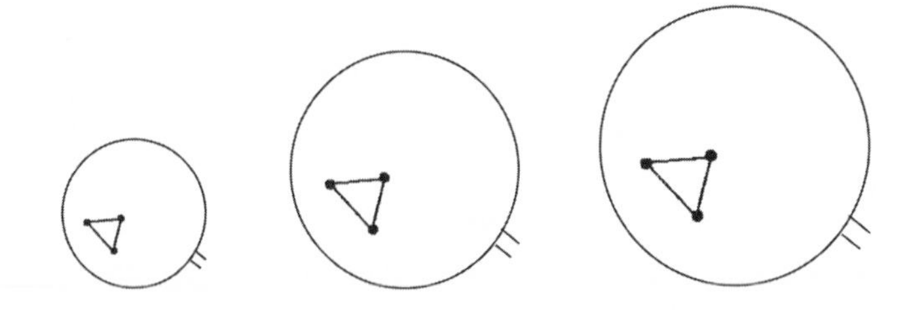

FIGURE 1.3 A BALLOON MODEL OF THE EXPANDING UNIVERSE

on the outer surface that represent clusters of galaxies, while the rubber between them represents the fabric of space. As we gradually inflate the balloon, the fabric stretches and moves the dots further and further apart. Something like that is happening to the cosmos.

But if the universe is expanding, what is it expanding from? How did it all begin?

We can calculate the rate of expansion, and then we can try and run the cosmic videotape backward, deflating the balloon as it were. The galaxies get progressively closer, and they move faster, so that as we get back to fourteen or so billion years ago, we find that the universe shrinks to a hot, dense mass, about the size of a grapefruit.

That is when scientists estimate the universe began, in what they call the big bang. In some type of cataclysmic eruption, time, space, and matter all came into being at once. It wasn't just matter expanding through space, but space itself expanding, taking the newly created matter along with it.

When the big bang theory was first proposed by George Gamow in 1945 on the basis of Einstein's theory of relativity, a lot of scientists didn't like it. One reason for their skepticism was that it sounded a bit too much like what the Christian church teaches about God creating everything from nothing. Many of them preferred Fred Hoyle's theory of continuous creation with no initial beginning. (To some, this theory seemed to avoid the religious problems associated with the big bang, although it isn't clear why continuous creation couldn't also imply a Creator.)

However, the big bang theory was beautifully confirmed in an accidental discovery by Penzias and Wilson in 1965. They found that the universe is bathed in a uniform wash of microwave radiation, cooled to just above absolute zero by expansion. This cosmic radiation is a remnant, an "echo" of the first second of the big bang.

What actually happened in the big bang?

Cosmologists have been working on this model for many years now, and are pretty sure of the fine details, even from the first millionth of a second. Alan Guth recalculated some of Einstein's equations and pro-

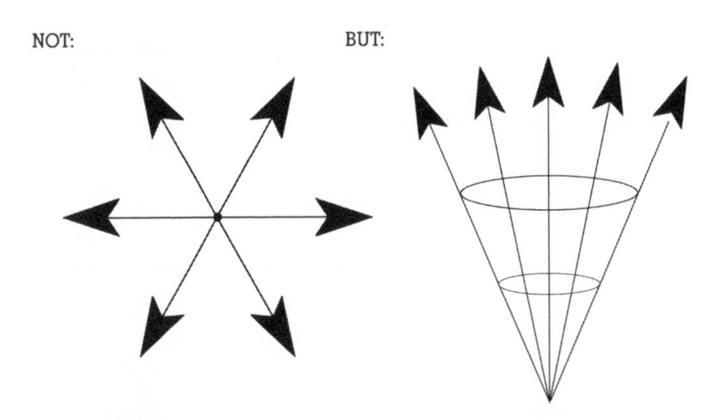

FIGURE 1.4 MATTER AND SPACE EXPANDING TOGETHER

posed that there had been a phase change in which the very early universe increased in size by 10^{30} in the first fraction of a second of time.[2] This inflation then ceased and the expansion as we know began, driven by the energy of an unthinkable explosion.[3] Then a decoupling of radiation and matter allowed atoms to form and led to the evolution of the modern universe—galaxies, stars, and ourselves. This model solved a number of problems and is now widely accepted.

There is no way of testing the inflation theory directly, but it does explain the overall smoothness of the universe and its so-called "flatness"—so finely balanced between continual expansion and ultimate collapse. When some findings that seem to confirm the theory were announced by NASA in 1992, a thousand astronomers burst into applause. What had been found were ripples from the edge of time, that is, irregularities in the radiation that pervades the entire universe, the afterglow of the big bang. NASA's Cosmic Background Explorer (COBE) allowed us to see the moment when the stars and galaxies of the universe first began to form—only 300,000 years after the big bang itself. But COBE hasn't given us all the answers; we still need to work out whether some cosmic repulsive force that works against gravity also played a role in the accelerated expansion.

What did the big bang explode into? What was around it?

When I tried to get an answer to this question from Arnold Wolfendale, then Astronomer Royal and professor of physics at Durham, his reply to me was, "You can't ask that question."

Physics has no answer: the laws of physics themselves break down at that point. There was no preexisting void into which the big bang happened.

There was no "before", unless one agrees with the philosopher and scientist Stanley Jaki that there may be higher embedding dimensions in an enlarged view of reality[4]—but the issue is not subject to proof either way.

It seems impossible for something like this just to happen, without any context.

Yes, it does run contrary to our intuition. For the universe to originate from nothing, as both Augustine and the big bang theory maintain, does seem to point to something outside it, something supernatural, as its source. There must have been some kind of creative force—a creator God whom many scientists have come to accept. When announcing the discovery of the ripples from the edge of time, the head of the COBE team, George Smoot, remarked that, "If you're religious, it's like looking into the face of God."[5] (Although, as Sir Bernard Lovell, astronomer and Christian commented, this is the God of technology, not of Christianity.[6])

If the universe is really expanding like a balloon, how long will it carry on expanding? Is the balloon ever going to pop?

Scientists are still looking for answers to this one. Here are the three possible scenarios they are working with. In the open scenario, the expansion accelerates more and more rapidly. In the critical scenario, the universe eventually reaches a critical point where the expansion steadies and stabilizes at this velocity, while in the closed scenario the universe slowly begins to deflate, like a balloon shriveling, until it reaches a critical point where it implodes in what some scientists call the big crunch.

When would that happen? We don't know. We do know that we seem to be near what is predicted to be the critical density. It remains to be

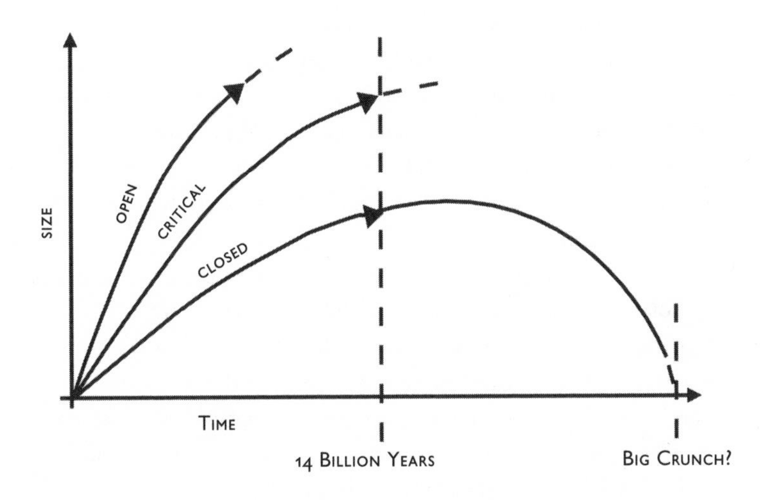

FIGURE 1.5 THREE SCENARIOS FOR THE EXPANDING UNIVERSE

seen whether the universe will then stabilize and continue to expand at a steady velocity, or, according to the current consensus, expand ever more rapidly, or start to contract and (perhaps quite suddenly) collapse. To be able to make more accurate predictions, scientists need to get a better understanding of dark matter.

Is there really such a thing as "dark matter"? It sounds like something out of* Star Wars*!

It's there all right. The galaxies are rotating much more slowly than they should, to judge from the visible star mass. Gravity must be affecting them in some way. But gravity from what? There must be something invisible out there, not made of ordinary atomic matter, that is slowing them down. And this invisible something must make up 90 percent of the cosmos!

Astronomers can't find nine-tenths of the matter in the universe! But they must be looking for it. What do they think it could be?

There are three possibilities: WIMPS, MACHOS, and Dark Blend (who says scientists have no sense of humor).

WIMPS stands for "weakly interacting massive particles," which may be neutrinos. Neutrinos are invisible, but billions of them are zipping through your body as you read this. Whether they actually weigh enough to make a difference remains a mystery. And there is some evidence of a supersymmetric particle called a neutralino, which may be a major component of what is called hot dark matter; another candidate is the axion.

MACHOS are "massive astronomical compact halo objects," such as black holes. Scientists call this type of thing cold dark matter.

Dark Blend is for those who want a bit of both. It's a scenario in which dark matter is a cocktail mix of hot and cold dark matter.

The best evidence yet of dark matter is the "bullet cluster"—the aftermath of the collision of two galaxy clusters. The gravitational mass far exceeds the luminous mass and proves dark matter exists—perhaps with a new fifth force—"dark energy" that influences only dark matter.[7]

I've heard about black holes—they suck in everything around them. But what are they? How did they form?

A black hole is a bit like a cosmic jail that imprisons both matter and light, but originally it was just an immense star. Every star is basically a nuclear plant operating at a temperature of several million degrees in which nuclear fusion converts leftover hydrogen from the big bang into helium. After 100 million years or so, a star has used up most of its hydrogen, and it starts to contract and burn hotter and hotter. As the fusion reaction heats up, atoms of heavy metals such as iron are produced. If the star is a relatively small one, like our sun, it eventually loses its heat and settles down to be a stable white dwarf. However, if the star is a large one, the pressure eventually becomes so great that the outer layers of the collapsing star explode in a fireball "supernova," like the one seen in 1987, outshining all the stars in the galaxy for a while. The force of the explosion flings the iron and other metals forged inside the star out into space, where they are captured by nearby stars and planets. This was the way our own earth acquired such a wide variety of minerals and metals—the source of all those components that now make up our own bodies!

After the supernova explosion, the remaining core of the star forms an immensely dense neutron star. But if the star was massive, the gravitational pull of the collapsed core is strong enough to pull everything around it into itself, including light—which makes a black hole completely invisible to any observer.

So how do we know they are there if we can't see them?

We can find them by looking for their effects. For example, we can look for binary or double stars. If one part of a binary pair has become a black hole capturing material from its companion, we would expect to see certain types of x-rays being emitted. Astronomers have found stars like this—one of them is called Cygnus-x.

The other place we would expect to find black holes is at the center of galactic discs. Colossal forces are at work there, including "quasars" (quasi-stellar objects), the most energetic objects in the universe. Astronomers have found evidence of the existence of black holes perhaps a billion times the mass of our sun in nearby (well, only a few million light-years away!) galaxies such as M51, M82, and M87. They suspect that a black hole of two or three million stellar masses is probably at the heart of our own galaxy.

NASA's Swift Satellite, launched in 2004, has so far found over 150 active galactic nuclei (black holes) out to a distance of 400 million light-years: the number is expected to rise to 450 within the next two years.[8]

What is there inside a black hole?
Is it just a whirlpool of trapped matter and light?

OK, here we have to go back to Einstein's work again. His famous equation $E=mc^2$ established that there is a close relationship between matter, energy, and time, and that each affects the other. He later went on to prove that matter actually distorts space, making it become curved, in the same way as the mass of the sun bends light passing near it from a distant star.

When we try to think about curves, we tend to think of them in three dimensions: depth, height, and length. But Einstein's work means that we have to include time as a fourth dimension that can also be curved in some way.

When a black hole forms, the enormous mass and energy crushed together at the center cause an acute curvature of space and a breakdown of space and time. It is as if the fabric of spacetime suffers a devastating rift and what is called a singularity is produced, a point with zero volume and infinite mass. No one knows what happens inside a singularity, although Roger Penrose and Stephen Hawking have proved mathematically that there must be one in every black hole.

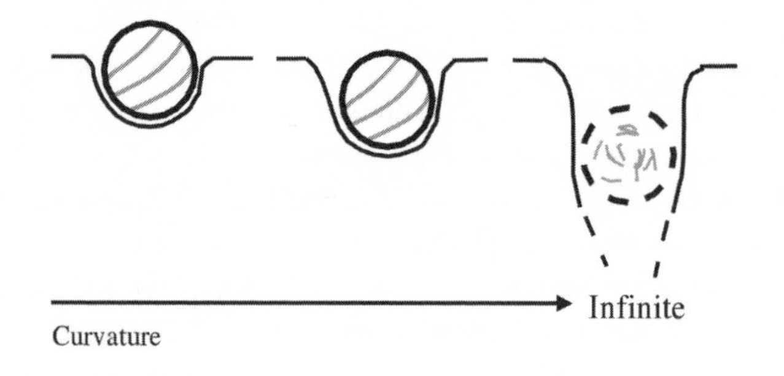

FIGURE 1.6 DIAGRAM SHOWING THE INCREASING CURVATURE OF SPACETIME

This is really hard to grasp, so let's try to use the balloon image again, but this time looking at one star on the surface of that balloon. The weight of that star will cause the rubber surface to bend or curve slightly around it. If the star is an enormously heavy neutron star with all its nuclei tightly packed together, it will bend space even more vigorously. The incredible concentration of mass in a black hole might even tear a small hole in the rubber . . .

How can there possibly be a tear or hole in three-dimensional space? And if there is a hole, can something fall through it? Where would it fall to?

Nobody knows. There are all sorts of theories about what could happen, many of which are very attractive to science-fiction writers, but scientists simply don't have any answers yet. It is possible that a hole could be an entry point to another universe altogether—a parallel universe perhaps with different dimensions. Or it might work like a wormhole, a tunnel through higher dimensional space, with the matter pouring out of what cosmologists call a white hole, the time reverse of a black hole, in some distant part of our own universe. Maybe we could one day use such shortcuts through the universe to reach another galaxy . . . Or, as Paul Davies suggests, something passing through a singularity could simply cease to exist in any form at all.[9] In fact, the whole physical universe could simply disappear in a massive singularity—although physicists such as Roger Penrose think this is unlikely.

But, you know, a lot of what you are saying sounds like a weird mix of science fiction and religion! Don't scientists find this troubling?

Of course they do, but they have to trust the calculations that lead them to these conclusions. When I wrote to Steven Weinberg asking how he wrapped his mind around the fact that Einstein's mathematics requires that spacetime be curved, he replied, "I don't imagine curved space; in this I let mathematics do my thinking for me."[10] Even Einstein, whose work laid the foundations for modern cosmology, didn't want to believe in black holes or an expanding universe. His general relativity equation includes a cosmological constant, which he added because he refused to accept his original calculations, preferring to believe that the universe was static and would go on forever. He later labeled this "my greatest mistake."

But if you think this is troubling, wait until you see what is happening at the other end of physics, in the weird world of quantum mechanics.

. . . . .

Of necessity, I had done most of the talking in this session, although the students had peppered me with questions. Now it was time to throw the ball back into their court.

So, what do you think of all this? What's your response?

There was a long pause while they tried to put their thoughts in order.

"It's very difficult to get a handle on this stuff," Melissa eventually said. "I'm an artist, and when I hear things explained, I try to form pictures in my head, but I can't get pictures for a lot of this. Your balloon image helped, but I am really out of my depth."

"It's not just because you are an artist, Mel," said Richard. "Ruth and I think we are scientists, and I don't think we did much better."

"Not to worry," I interjected, "we just saw how even cosmologists like Einstein have trouble thinking about this."

"What did you make of it, Angus?"

"I found the sheer size of the cosmos mind-blowing. I can hardly imagine how far it is to a star, still less to a black hole at the center of our galaxy. It is awesome—'out of this world!' And all this talk of the universe coming out of nothing and of rifts in spacetime makes modern cosmology sound like a mix of sci-fi and religion . . . It isn't the sort of science Richard was talking about when he said he wanted to concentrate on what we can touch, see, and visualize, and on experiments we can repeat."

"Yes," admitted Richard, "it seems that here we can't prove anything directly. All we've got to do is have faith in the mathematical models based entirely on theoretical equations."

"Faith in mathematics is OK as far as it goes," interrupted Angus. "But what about the places where mathematics doesn't go? Before the big bang? Inside a singularity? It doesn't seem to me that this lot rules out forces that we know nothing about, and which might be related to those mysterious feelings we talked about at the stone circle."

"Oh, that's space aliens at work," laughed Ruth.

"And where do those aliens come from? Other galaxies? But those galaxies also didn't exist till that orange-sized ball of something exploded for some reason. Where did that ball come from? Just adding distance doesn't solve the problem completely. Is there a creator of some kind out there somewhere?"

"Yes," said Mel, "I find the distances and the size almost frightening. It's all so big. And I am so small. Yet when I look at my runes or the Ouija board I expect these mysterious forces in the universe to take notice of my stupid questions! Am I nuts?"

"OK. That's the bell for the end of the first round," I broke in. "We'll discuss Mel's sanity later! All of us are a bit overwhelmed by the huge scale of this universe, so next time we meet, we'll go the other extreme, and start to look at the things that are right at the other end of the scale."

2. Mystery, Models, and Quantum Theory

Anyone who is not shocked by the quantum theory has not understood it. —NIELS BOHR[1]

IN OUR FIRST SESSION, we had talked about the huge scale of this universe. This time we were heading in the opposite direction, looking at physics on the atomic level. Once again, the group had plenty of questions—and my answers generated even more.

Who first dreamt up the idea that atoms exist?

The ancient Greeks were the first people to talk about atoms as the smallest indivisible particles of matter, too small ever to be seen. However it took more than 1,900 years for other scientists to agree with them and to decide that atoms were probably like billiard balls. Physicists agreed that Newton's laws governed their behavior and that Maxwell's equations explained the electromagnetic phenomena associated with them. Barring a few minor puzzles still to be solved, we understood the physics of our world. And then between 1895 and 1926 this whole structure came tumbling down like a pack of cards.

What happened? Newton's laws didn't stop operating.

No, it wasn't those laws that changed. It just suddenly became clear that those laws only applied to a tiny aspect of reality, and that things were much more complicated than we had thought.

The first inkling that atoms were not solid came with Becquerel's discovery of radioactivity in 1896. If atoms were emitting something, they were obviously more than just billiard balls.

Then in 1900 Max Planck calculated a formula to explain the radiation from what scientists call a blackbody, an ideal object that both absorbs and then emits all the radiation falling on it. Planck found that the only way he could calculate the radiation was to assume that the energy radiated by atoms was not infinitely variable but could only change in definite amounts, which he called "quanta."

Einstein worked with this idea, and found that light, which scientists had assumed was like a wave, also behaved like a quantum, a particle. His theory of the photoelectric effect implied a type of interaction between light and matter that classical physics couldn't account for. Wrestling with this and other problems led Einstein to develop his special theory of relativity, in which space and time were fused in one four-dimensional continuum.

But what has all this got to do with atoms?

Once scientists had got past the billiard ball model of an atom, they needed to find a better model. In 1912 Ernest Rutherford experimented with firing alpha particles (positively charged helium nuclei) at thin layers of gold. When some of his alpha particles came flying right back at him, he realized that they must have been repelled by something that also had a positive charge, and he identified this as the center or nucleus of each atom. He then developed a model of the interior of the atom that was based on the planetary system—at the center was the sun (the nucleus) made up of neutrons and positively charged protons around which the negatively charged electrons orbited like planets. Niels Bohr pointed out that if it was assumed that the orbits of the electrons were fixed, this would help explain the quanta which Planck and Einstein had found. A fixed packet of energy would be released each time an electron jumped from one orbit to another.

While the Rutherford-Bohr model is no longer accepted, it did make the important point, which we all keep forgetting, that everything consists mainly of nothing!

What do you mean "everything consists mainly of nothing"?

That's just one of the paradoxes that quantum theory is full of. If you think of the nucleus as the size of your thumbnail, the rest of the atom,

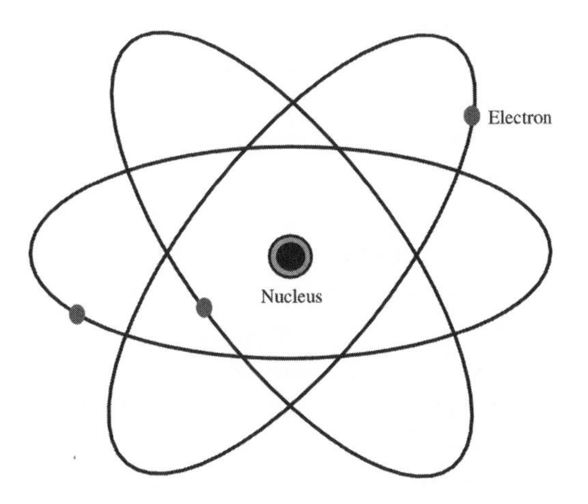

FIGURE 2.1 RUTHERFORD-BOHR MODEL OF THE ATOM

the part where the tiny electrons are orbiting, is the size of a football field or of St. Paul's Cathedral. Most of the atom consists of empty space.

So take that chair you are sitting on: it looks pretty solid to me now. But as a scientist, I know that you are actually leaning back on a lot of empty space, with a few odd bits of atomic particles here and there held together by electric force fields. And of course the same applies to the body that is doing the sitting—and to your brains that are boggling at this idea!

You said that was just one of the paradoxes of quantum theory. But what exactly is quantum theory?

You remember that early in the twentieth century, scientists found that light behaves both like a wave and like a particle—although it seems logically impossible for it to be both at once. The word Planck used for this particle was "quantum," and so quantum theory is the study of physics at the subatomic level, where there seem to be a lot of things that are confused about their identity. (A confusion that they communicate to physicists . . . Roger Penrose has pointed out that "Many phenomena in the microcosmos of quantum mechanics fly in the face of common sense"!)[2]

But quantum mechanics clearly works: it has given us lasers, transistors, superconductors, electron microscopes, and atomic power. So we have to believe the mathematics, despite the fact that it often means we have to abandon what seems like common sense and accept seemingly absurd or miraculous events.

That sounds weird—but I don't really know what you're talking about. Can you give us some examples?

OK. Let's start off by looking at probability.

In the 1920s Erwin Schrödinger set out to produce a mathematical model that would describe the mechanics of an atom, in the same way as Newton's physics had described the mechanics of our solar system. He found the best way to do this was to treat particles of matter as if they were waves and developed his famous wave equation, which is fundamental to almost all modern chemistry and physics. This equation is really only a description of the probability of finding an electron somewhere. Nothing in the quantum world is ever certain—except uncertainty!

Heisenberg focused on this built-in unpredictability when he developed his uncertainty principle, which shows that you can never know both the position *and* the speed of a particle. If you know where a particle is, you don't know how fast it's moving, and vice versa.

Why? Because to find out where it is, you have to stop it moving, and if you want to know how fast it is moving, it won't stay in one place. As soon as you start to measure something, you change the thing you are measuring.

But all the science we did in school was based on predictions and measurement. Does that all go out the window?

Not completely. The classical physics taught in schools is based on nineteenth-century thinking that was dominated by the idea of determinism, the belief that things were always linked in a long chain of cause and effect. That idea is still useful for normal large-scale experiments. But it's no use on the atomic level where probability and uncertainty are now fundamental concepts.

Classical science also liked to be able to build models of things, so that we could visualize them. (Didn't some of you build models of the solar system, or of atoms and molecules for school science projects?) But you simply can't visualize quantum mechanics. How can you build a model that includes uncertainty? Or construct a particle that is also a wave?

This wave/particle stuff seems to be very important, because it keeps coming up. Can you explain it a bit more?

The fact that particles can be described as wave functions is one of the key paradoxes, although probably we would be better off describing it as a mind-blowing "double truth" rather than a paradox.

To give you just one example of this bizarre wave/particle duality, take the double-slit experiment. Suppose you were to shine a light through two slits in a sheet of metal. The pattern you would see on a screen on the far side would show an interference pattern, with light from the two sources overlapping, just as waves overlap if you throw two stones into a pool. Now suppose you had very fancy equipment, and were able to emit exactly one photon (one quantum) from your light? What would you see? Would there be an interference pattern or not?

There shouldn't be any interference. If there really is only one photon, it can only go through one of the slits, right? And it would produce a neat dot on the screen, proving it's a particle.

That's what you'd expect. And if you fired off a succession of photons, you'd expect to get two small sets of dots, depending on which slit each photon went through. But when scientists did this experiment and looked at the pattern gradually building up on the screen, what they saw was not two separate sets of dots, but the same interference pattern they would have obtained if they had used a larger light source.

Does that mean that the photons broke up as they went through the screen, so that some parts went through one slit and some through the other? Couldn't you put some sort of tag on the photon, so you could see what it was doing?

Not really. You remember what I said about uncertainty and the Heisenberg principle? Until it hits a screen the photon isn't a "thing" you can tag or split. It's a probability wave function. And it stays a wave function until it hits the screen and the observation is made. At that moment, the wave function collapses, probability becomes concrete reality, and we know where the particle is. We had no way of knowing where it was before we observed the dot on the screen. And that problem of how to find the relationship between the wave function and the observed event is one of the major problems for scientists today.

If a photon isn't something you can split, how can there possibly be an interference pattern, which involves scattering of light? It seems as if one photon must be passing through both slits at the same time!

That's the whole problem of quantum mechanics—we don't know the answer. Erwin Schrödinger was keenly aware of this problem, and used the example of a cat to illustrate it. Suppose you take a cat and put it in a box with an atomic clock that may or may not trigger a switch that will break a phial containing a lethal dose of cyanide. The box is sealed, so you can't tell what is going on inside. Like the quantum that went through two slits at once, quantum mechanics says the atom may be in both states at once: that is, it may be in the trigger state and the nontrigger state. So is the cat alive or dead—or is it both?

If it's both, I presume you have to have parallel universes, in one of which it's alive while it's dead in the other.

Some physicists have actually suggested this! Weirder still, it has been suggested that in addition to our world there are an infinite number of parallel universes in which all other possible results of actions appear. Every time there is more than one possibility, a new universe splits off—unconnected with every other universe. However, this many-worlds theory of Everett and Wheeler has now been abandoned, even by John Wheeler himself, because it "contains too much metaphysical baggage."[3]

Another way out has been suggested—that all the possibilities in the probability wave function are "summed over history," to give only *one* observed reality. This too is now largely abandoned, as it seems to bend the rules too much. Other physicists argue in favor of decoherence, a

FIGURE 2.2 SCHRÖDINGER'S CAT

type of censorship of possible alternative worlds in which the environment seems to act as an observer, collapsing the wave function. Physicists themselves are divided. The mystery remains.

Surely you could solve the whole problem just by lifting the lid to looking at the poor cat? Why bother to imagine other universes when you can simply observe what has actually happened?

This was part of the point Schrödinger was making when he used this example. But what is called the Copenhagen interpretation of quantum mechanics (supported by Bohr in 1926 and accepted by many physicists) says that the uncertainty is a fundamental principle, and that as soon as you open the lid, you collapse the wave function and create a reality that didn't exist before.

Huh? What do you mean "collapse the wave function"?

Remember Heisenberg's uncertainty principle? One of its implications is that as soon as you observe something, you change it. As soon as the photon hits the screen, it stops being a wave. In the same way, as soon as you actually see the cat, the 50–50 probability that the cat is alive or dead is settled one way or the other. You create that reality by opening the box.

Are you saying that if there weren't somebody to observe an event, it wouldn't happen? This sounds a bit like the question "If a tree falls in the forest and there is nobody around, does it make a sound?"

Yes, the orthodox Copenhagen interpretation says we ourselves are involved in the nature of reality in a fundamental way. We can no longer talk about absolute scientific objectivity, because consciousness plays an essential role in the nature of physical reality. What we observe is all that we can know.

Do all physicists accept this?

Schrödinger didn't like it; nor did Einstein. Although Einstein's own discovery of the photoelectric effect paved the way for the idea that energy only comes in discrete quantum packets, he was never able to accept the elements of uncertainty and indeterminacy in quantum theory: "God does not play dice with the universe" was his famous exclamation. "Laws which compel the Good Lord to throw dice in each individual case, I find highly disagreeable."[4]

Working with two other physicists, Podolsky and Rosen, Einstein set out to demolish the quantum worldview by exposing its absurdity. He attempted to challenge Bohr's interpretation in his 1935 "thought experiment" in which he examined the decay of an unstable elementary particle into two photons, which were then allowed to fly apart to opposite ends of the universe. As soon as we measure the spin of one of the particles, said Bohr, we know the spin of the other. Because measuring affects the spin of the particle, as soon as we measure one particle, we also affect the spin of the other, no matter how far away it is. Bohr argued that this doesn't just apply in the microscopic realm, but even in objects large enough to be seen with the naked eye.

Einstein scoffed at this type of "spooky action at a distance," but theoretical work by John Bell in the 1960s demonstrated the inevitability of nonlocal influences, and his theory was confirmed in an experiment by

Alain Aspect in the early 1980s, overturning Einstein's view. Aspect's findings could not be explained without action taking place at a distance, however "spooky" this action might be.

However, if you talk to physicists, you'll find that many of them don't think these things are particularly bizarre. The Copenhagen interpretation helps them to avoid many of problems by insisting the problems are meaningless, sweeping the paradoxes or double truths under the carpet. But as Murray Gell-Mann famously remarked in his Nobel Prize lecture in Sweden, "Niels Bohr has brainwashed a whole generation of theorists into thinking that the job was done fifty years ago."[5]

Many physicists live with the paradoxes by acting as if they still believe in real probability waves or billiard ball–type particles. In most cases, they don't worry too much about philosophical issues until they start to develop a quantum theory of the universe. But as soon as you start working with big bang theory, you find that when the universe was truly microscopic, Einstein's equations break down and quantum events become very important.

If quantum events require an observer to be real, who was the observer at the big bang? Or isn't that a question that should be asked?

It is a valid question if we accept observer-centered reality, but it's one we have often been scared to ask because we have been so afraid of introducing any idea of God into physics. Of course, it is right to avoid merely using God as an explanation to fill the gaps in our own knowledge, but we don't have to eliminate the possibility of a Creator God completely from our thinking.

There are a pair of limericks that summarize the problem quite neatly. The first is by Ronald Knox:

> There once was a young man who said "God
> Must think it exceedingly odd,
> If he finds that this tree
> Continues to be
> When there's no one about in the Quad."

To which Bishop Berkeley could have replied in a limerick letter:

Dear Sir,
Your astonishment's odd,
I am always about in the Quad,
And that's why the tree
Will continue to be
Since observed by,
Yours faithfully,
GOD.[6]

Are you suggesting that an "observer-centered reality" must include God?
It's certainly not incompatible with the existence of a God, is it? It certainly appears that the whole idea of what is real is a lot more complicated than we thought when we first talked about science.

. . . . .

Once again it was time to throw the ball back to the students.

So, what do you think of all this? What's your response?
This time, the pause was even longer before Ruth admitted, "I find it very difficult to understand this. I'm prepared to believe it works, but I haven't a clue how."

"You're in good company," I said, "even excluding the company here who all agree with you wholeheartedly. I like the comments by Richard Feynman, who won the Nobel Prize for physics: 'Screwy . . . a mathematical absurdity—like one hand clapping . . . Anyone who isn't taken aback, hasn't appreciated the mystery at the heart of the universe.'"[7]

"But I don't like mysteries," said Richard. "Mysteries are Angus's things, and Mel's. I want science I can get my teeth into, that makes sense. This sort of stuff is starting to sound altogether too religious for me."

"What do you mean 'too religious'?" I probed.

"Well, partly I don't like the idea of God as observer, creating reality. But that's not all. It's religious in another way too."

"Is it the fact that we have to take a lot of this on trust that's the problem?" asked Ruth.

"No, or only partly. There are a lot of things I have to take on trust because I haven't got time or the ability to check the facts . . ."

"STOP right there," said Angus. "We have just learned that there aren't any facts! AARRGGGGGGG!"

"OK . . . Let's put that one on hold for the moment," said Ruth. "I think this lot is a bit like religion because it means you have to believe things that sound contradictory and illogical. I suppose the wave-particle stuff is a bit like Christians saying that Jesus was both God and human?"

"Yes," said Mel excitedly, "and the stuff about things having effects at a distance that Einstein didn't like. Could that tie in with my question last week about the universe being too big to notice me? Maybe things that are a long way away do affect the answers I get when I look at my cards or my board?"

"Well," I said, "certainly the fact that things can have an effect at a distance might mean that prayer isn't as way out an activity as we might have thought. Roger Penrose has pointed out that if the quantum inseparability or nonlocality effect were to operate on a macroscopic scale, it would fly in the face of common sense.[8] We need some more complete but as yet undiscovered theory.

"You aren't the only ones to have noticed the religious type of language that physics slips into here. Books called *The Tao of Physics* and *The Dancing Wu Li Masters*[9] tried to combine physics with Eastern mysticism. It seems that a new type of language is needed across many areas of physics to describe such transcendent models and to cope with the paradoxes of the quantum theory.[10]

"But it's not only a new language that is needed. One of the major problems physicists have faced is that the two main theories of twentieth-century physics, general relativity and quantum mechanics, seem to contradict each other. So scientists have been looking for a new framework, and think they have found it in what are called "superstrings." If you think that what we have looked at so far is hard to handle, wait till next week when we'll explore the fact that physics is giving us a universe with more dimensions than meets the eye."

3. Quarks, Superstrings, and M-branes: A Theory of Everything for the Twenty-first Century?

If we do discover a complete theory, it should in time be understandable in broad principle by everyone . . . Then we shall all, philosophers, scientists and just ordinary people, be able to take part in the discussion of why it is that we and the universe exist. —STEPHEN HAWKING[1]

IN OUR LAST SESSION, we looked at the fact that atoms have been variously described as billiard balls, as miniature solar systems, as nuclei surrounded by a fuzzy cloud of electrons, and as waves in multidimensional space. For a time, physicists tried to ignore the problem of having to treat a particle such as a proton as a point with no dimensions. But some of them began to search for other solutions, for a "theory of everything" that would explain the contradictions they were being forced to live with.

So did they develop a new model of the atom?

In 1964 two physicists, Murray Gell-Mann and George Zweig, developed a beautifully simple model that could explain all the known particles. It is based on the idea that the protons and neutrons in the nucleus of the atom are not solid units, but are themselves built up of smaller particles. Gell-Mann and Zweig called these fundamental building bricks of matter "quarks" after the line "Three quarks for Muster Mark" in *Finnegan's Wake* by James Joyce. (Mel, was it you who said that scientists were an illiterate bunch?)

But what they constructed was just a model. Do we have any proof that quarks actually exist?

At first all they had was a mathematical, nonvisualizable, and untestable idea, which was ridiculed by professors of physics. But within ten years, to Gell-Mann's own surprise, the model became more and more useful and was accepted as part of the standard model.

Yet no one has ever seen a quark, and there is still no direct experimental proof that they exist. They never seem to occur as ordinary free particles. But when scientists make predictions based on the assumption that they are there, and conduct experiments to test these predictions, the results are what they were expecting. You might say that "the proof of the pudding is in the eating," and this particular pudding tastes strongly of quarks.

What are quarks like?

There seems to be six different kinds of quark, which particle physicists describe using terms such as up, down, strangeness, charm, top (or truth), and bottom (or beauty). The top quark was only discovered in 1995. These properties, which give quarks different "flavors," are best thought of as degrees of freedom or extra dimensions rather than as ordinary measurable properties. It also seems that each quark can also exist in three colors, another tongue-in-cheek label for a nonmeasurable property. To put this in terms of a very simple image, you could say that quarks are like a mathematical ice cream that comes in six different flavors with three different toppings!

You said that the new model claims that protons and neutrons are made up of quarks. But what about electrons? Where do they fit in?

Electrons are now seen as a subset of a larger category called leptons. So in terms of this model, the world as we know it is made up entirely of quarks and leptons and the interactions between them.

As far as we can tell, these interactions are governed by what scientists have identified as the four basic forces of nature: gravity, electromagnetism, the strong force in the nucleus, and the weak force of radioactivity. In current theory these four are transmitted by the exchange of messenger particles; for example, electromagnetism is carried by photons.

But we are still trying to understand these forces better. After all, we haven't known about some of them for very long. It was only in the nineteenth century that Maxwell realized that the apparently different phenomena of electricity and magnetism were really the same force, which we now call electromagnetism. By the time of Einstein, only the forces of electromagnetism and gravity were known. So the other two forces are really the new kids on the block, and we still have a lot to learn about them and about their interactions with the other forces.

One thing we have discovered is that there is a deep connection between the electromagnetic field and the weak nuclear force. Abdus Salaam, Steven Weinberg, and Sheldon Lee Glashow made this discovery of the unified "electroweak" theory independently in 1967. In 1974 Glashow, together with his Harvard colleague Howard Georgi, suggested an analogous connection might be made with the strong force. They produced what is called the grand unified theory of three of the four forces (GUT). But what physicists really want is a theory of everything that will explain how all four forces are linked together.

Would a theory of everything also deal with the problem of waves vs. particles that we talked about last time?

Quantum field theorists put a lot of work into trying to solve that problem, focusing particularly on the apparent conflict between general relativity (Einstein's theory of gravity) and quantum mechanics. Most of their attempts were not very successful. Then, in what is called the September 1984 Revolution, Michael Green and John Schwarz introduced the idea that particles are not points after all, but strings—in fact ten-dimensional superstrings.

The idea of strings had been used in the late 1960s to describe the glue holding quarks together, but there had been a lot of problems with this theory. However, this new approach solved these problems—and pointed to a possible complete unification. Superstring theory actually contains Einstein's gravitational theory within itself in order to be consistent and avoids the problems of point particles. At low energies, tension pulls the strings into points, but their essential "stringiness" becomes evident at high energies.

Green and Schwarz's original attempt to use string theory required twenty-six dimensions. This was at first dismissed as unrealistic ("the model has a sickness," as the Norwegian Holger Nielsen poetically described it to me). Everyone "knows" there are only three space dimensions and one time dimension, so their work was revised and a much better answer was reached: only ten dimensions are necessary, nine for space and one for time.

Big deal! They solved the problem of things being both particles and waves by adding an extra six dimensions! Some solution!

Scientists don't like it either, but the only reason the idea of superstrings has been accepted is that it works and is enormously creative. Ed Witten, one of the foremost exponents of current theories, claims that "all the really great ideas in physics are spin-offs of superstrings—even General Relativity" and that discovering this was "the greatest thrill of my life."[2]

No one has fathomed the full extent of this theory, but Professor Witten and other key physicists today are confident that just as the vibrations of a violin string give rise to different notes, so the vibrations of superstrings give rise to all the different particles of matter and all the forces acting on them.

But where did the idea of these extra dimensions come from? How did it even begin to be thought of?

The first physicist to attempt to introduce five dimensions into our description of spacetime was ignored until this last decade or so. He was an obscure German academic called Theodor Kaluza who was trying to unify gravity and electromagnetism. Kaluza's son remembers the exact moment of inspiration. He was eight years old at the time, and was reading quietly in his father's study. Suddenly his father stopped writing, "was still for several seconds, whistled sharply and banged the table. 'It works!' He stood up, motionless for a while—then hummed the aria of the last movement of Mozart's *Figaro*."[3]

Kaluza had found that the equations for gravity and electromagnetism were identical—if five dimensions were used and not the usual four. He excitedly wrote to Einstein in April 1919 to enlist his support for this idea, and received a postcard reply: "The thought of achieving this unification through a five-dimensional world has never occurred to me and may be completely new. Your idea is extremely pleasing to me."

However, in his subsequent letters, Einstein discouraged Kaluza, first with one and then another mathematical quibble, or by saying his paper was too long for publication, or by suggesting he "try another journal." It was two years before Einstein finally wrote that he was having second thoughts about "having restrained you from publishing your idea on the unification of gravitation and electricity."[4] This postcard was framed above the mantelpiece in Kaluza's son's home in Hanover, where I sometimes used to visit him. Kaluza was overjoyed at Einstein's encouragement, and published his paper. But it was so far ahead of its time that it caused little stir. Oskar Klein, a Swedish physicist, attempted some work in quantum mechanics using five dimensions in 1926, and produced the Kaluza-Klein model. Louis De Broglie considered linking it with his own idea of pilot waves, guiding waves directing each three-dimensional particle from outside spacetime, but that idea died, and the modest and unassuming Kaluza himself died in 1954, without ever achieving the recognition he deserved. The challenge of this fifth dimension that couldn't be visualized needed more mathematical tools—and a new world picture that only emerged in the 1980s.[5]

Today, however, the Kaluza-Klein model is widely used. The concepts of supergravity, strings, and superstrings have led to an appreciation of the idea of extra dimensions—and not just five dimensions, but ten of them.

But where are these extra dimensions?
How can we visualize them?

In modern physics, these extra dimensions are often treated as physical things, really existing, and not just as a mathematical device. But it is difficult to visualize them. One of the original discoverers, Michael Green, described them to me as a spiral at each point in space. All the dimensions are present at each point, but we are only aware of our three dimensions plus time (the external dimensions) because all the others

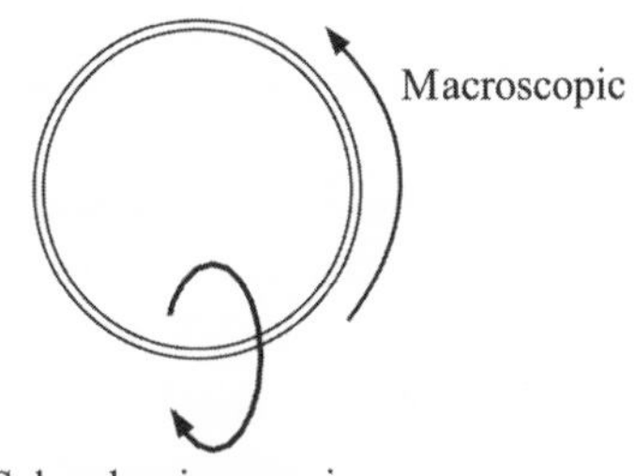

FIGURE 3.1 INTERNAL AND EXTERNAL DIMENSIONS OF REALITY

(the internal dimensions) are curled up or compactified into a tiny space (10^{-30} cm) and are too small to notice.

Alan Chodos and Steven Detweiler of Yale University suggest another image:[6] picture a very long, very thin piece of tubing that has been joined on itself to form a closed ring. The small circumference of the tube itself is the internal dimension, while the larger distance around the perimeter of the circle is the macroscopic distances we actually observe.

Chodos and Detweiler also happily point out that it is quite possible that the internal dimensions may eventually expand again.[7] Brian Greene also suggests that it is quite possible that one or more of these extra dimensions may well be comparatively large, or even infinite! We still have a lot to learn about them.

But do physicists really believe that these dimensions exist?

The jury is still out. Many physicists—among them Paul Davies, John Schwarz, Edward Witten, Michael Duff, and Michael Green—take the Kaluza-Klein dimensions very seriously, suggesting that our four physical dimensions are perhaps shadows or projections of a higher dimensional reality. But for others, like Richard Feynman, the lack of testable predictions remains a problem.

However, it works! The two major theories of this century—gravity and quantum mechanics—are incomplete. Superstring theory resolves the differences between them by using additional dimensions of spacetime. But a further problem is that there are five competing versions of string theory.

Have they yet decided which one of the five works best?

Actually, as you might expect by this stage, all of them do! Just as Maxwell found that electricity and magnetism were aspects of the same force, so in the late 1990s, a new theory, called M-theory, emerged that finds a simple mathematical relation between several quite different theories of superstrings. It does this by identifying an amazing "duality," a mysterious symmetry whereby the five theoretical models only appear to be different, but are actually part of one single underlying theory, describing exactly the same physics. We've seen this sort of duality before, in the way light is both a particle and a wave. On a more mundane level, we've even seen it in the way both ice and steam are forms of water.

The discovery that duality also applied to strings was so startling that it has been called the Second Superstring Revolution of 1995. In a mathematical tour de force, Ed Witten drew together all the work on dualities under the umbrella of M-theory in eleven dimensions. He insists that the "M" can stand for magic, mystery, marvel, membrane, or mother of all strings! His audience was stunned by Witten's evidence for this new duality, which Paul Townsend later described as "a democracy of branes," in which all brane theories were equally valid.[8]

What exactly is a membrane?

M-theory replaces the rubber band–like loops of superstrings with something more like soap bubbles, called supermembranes. Starting from 1-D strings and 2-D membranes, physicists add on dimensions, so that Kaluza's idea would be regarded as a 5-brane in physics-speak. When there are "p" dimensions, they speak of p-branes (and yes, they are perfectly well aware of how this is pronounced!).

If they use a general term like "p-branes," does that mean that they aren't certain how many dimensions there actually are?

Yes, it does. Cumrun Vafa of Harvard University thinks that there may even be an all-encompassing theory in twelve dimensions, which he calls the father theory (F-theory). Intriguingly, the extra dimension required here is another one of time, rather than space. Schwarz thinks there may be even more dimensions, and Green and Townsend think there may not be any fixed number of dimensions at all.

Witten and Vafa suspect that we will see another superstring revolution before agreement is reached. Then in 2003 Edward Witten unlocked the secret of combining superstrings with the revolutionary but inherently different six-dimensional twistor space, developed from Roger Penrose's twistor theory of the 1960s—both candidates for a theory of everything. More recently, Professor Lisa Randall of MIT and Harvard created headlines, "Why I believe in higher dimensions!"[9]

In our first session, we talked about black holes and alternate universes. These invisible dimensions sound a bit similar. Are there any links between these ideas?

There is a special class of p-branes called D-branes (from Dirichlet), which provide "edges" to spacetime. They may throw light on the conjecture that an object may pass through a black hole and enter another universe—a universe with different dimensions of spacetime. But we really don't know enough yet to do more than speculate.

These multidimensional branes may also link up with John Wheeler and Kip Thorne's idea of wormholes, tunnels in the geometry of space and time, connecting otherwise distant or completely disconnected regions of the universe.[10]

We also talked about the big bang— how does membrane theory fit in with that?

Many scientists believe that in the beginning there were (at least) ten dimensions. At very high energies, such as at the start of the big bang, all these dimensions were equal, but their symmetry seems to have been broken in a phase transition that occurred 10^{-43} seconds after the big bang. Six of the ten dimensions then curled up at each point in space, which made the superstrings look like particles. Meanwhile, the four dimensions that were to become our physical space and time expanded. Interestingly, mathematicians point out that it could equally have gone the other way, with three dimensions compactifying and six (plus the one time dimension) expanding.

They say all this, but they haven't yet got any direct evidence for it, other than the mathematics. Is there any way they can hope to prove any of it?

The way physicists try to prove things is by making predictions about what should be true if this theory is true, and then looking for evidence that their predictions match reality. One of the predictions based on superstring theories is that there are superpartners—that nature is concealing a companion particle for every basic particle now known. The partners of quarks are called squarks, those of electrons are selectrons, and so on. These superpartners must exist if the universe is governed by supersymmetry, a way of relating the apparently different appearances of matter and forces.

Physicists are actively engaged in looking for these, and are hoping to find them. To help them, they are using tools like the new generation of particle accelerators such as the LHC (Large Hadron Collider), which will begin operating in the CERN laboratory in Switzerland in November 2007. This is the world's most powerful particle accelerator, combined with the ATLAS detector. It will analyze the debris of the quantum "shrapnel" produced by slamming protons into one another. By using computers to reconstruct the signature combinations produced, scientists hope to find evidence of extra dimensions, supersymmetry, and even of dark matter. They can then claim the discovery of new particles—even though they haven't observed them directly.

As Michio Kaku reminds us, although it remains to be confirmed directly, string theory "has no rival as a theory of everything" in a unified theory.[11]

It seems that, unless I can do fantastic mathematics, I have got to take an awful lot of this on faith. And even if I could do the mathematics, I just have to believe the numbers, with no experimental facts to back me up. This is all still just a beautiful idea that somebody dreamt up! It sounds almost like religion!

You're right, and one of the winners of the 1979 Nobel Prize for physics backs you up. Sheldon L. Glashow wrote that for the first time since the Dark Ages "we can see how our noble search may end, with faith replacing science once again!"[12] In fact a generation of physicists have devoted their careers to string theory without getting any experimental feedback. They risk investing a lifetime of effort for an inconclusive result.

But surely if they can come up with this "theory of everything" that Hawking talks about, it will all start to fall into place neatly?
I wouldn't bank on that. It's a bit of an exaggeration to call it a theory of everything. All it will do, if and when they get it sorted out, is give us a theory about how the physical universe operates. It doesn't answer other questions, about us, our very existence, or why we are here. Even a militant atheist like Nobel physicist Steven Weinberg admits there are mysteries science will never be able to resolve.

But you know what really gets me is the feeling that I can't trust my senses any more. None of this makes sense. Reality—us, sitting here on chairs—doesn't seem quite "real" enough any more. There are all these other dimensions floating around, and real scientists seriously talking about alternate universes.
I know what you mean. But did you know that the whole idea of not being sure about what reality is has a very long history. Next time we meet, I'll introduce you to Plato and his cave. Way back in about 400 BC he was already speculating that we may be only the three-dimensional shadow of a higher dimensional universe, which sounds very much like the position modern physicists are adopting on the basis of their elegant mathematics.

4. What Is Reality?

The entire world which we apprehend through our senses is no more than a tiny fragment in the vastness of Nature. —MAX PLANCK[1]

OUR BRIEF SURVEY of current thinking by cosmologists and quantum physicists had emphasized that much of what we can prove mathematically runs completely counter to our commonsense view of the world. Paradoxes abound: a particle is also a wave; an observer creates what is observed; we can only experience four dimensions, but there are a further six that we cannot see.

The group's notions about the certainties of science had been shaken. While they were prepared to believe the mathematics, they still needed to come to terms with the fact that "reality" was less real than they had thought. There was a temptation to dismiss the multiple dimensions of modern physics: "If we can't see them, they don't exist," or "If we can't see them, they don't matter." But the overriding desire was to find a way to think about these concepts that made sense of what they had learnt in relation to their own lives.

What they were in fact looking for was a "metaphysic" in its ancient meaning of something that comes after the physics—which inevitably led us beyond mathematics and into the realm of philosophy.

You mentioned that Plato said something about living with multiple dimensions. What on earth did he know about it?

Although Plato lived about 2,400 years ago, he was acutely aware that what we see is not all that there is to see.

Plato found that the best way to explain this was to construct a picture that people could keep in their minds. His illustration worked so well that to this day almost everybody who ever does a course in philosophy is told about the allegory of the cave.

Imagine a chamber like a cave with an entrance open to the daylight, but running a long way underground. In this chamber are men who have been imprisoned since childhood. Their legs and necks are fastened in such a way that they cannot turn around to see what lies behind them. Behind them a fire is burning, and between the prisoners and the fire runs a road along which people pass carrying all sorts of objects and chatting to each other. All that the prisoners can see are the shadows of these people in the shapes that the light of the fire throws on the wall of the cave opposite them. The prisoners can see these shadows moving; and when they hear talking, they assume that the shadows are talking, for the wall reflects the sound from behind them. They believe that the shadows of the objects are the real thing.

Suppose one of the prisoners were released and turned around to look directly at the real objects and people: he would be too dazzled to see them properly. If he were told that the shadows that he used to see were mere illusions and that he was now nearer reality and seeing more correctly, would he believe this? Wouldn't he think that what he used to see was more real than the strange things he is now seeing?

But suppose they took him out of the cave, into broad daylight—wouldn't that convince him?

But he would be even more dazzled out there. At first he would probably much prefer just to look at shadows on the ground; after a while, he might be able to handle looking at reflections of men and other objects in water; and it would probably be quite some time before he could actually look at the objects themselves. Even then, he would probably prefer to look at things at night, by the light of the moon and stars, rather than in the full glare of at the sun.

But once he had managed it, wouldn't he be so excited that he would want to rush back in and haul out all the other prisoners, so that they could see what he had discovered?

Yes, but he might have some difficulty convincing them to move.

After all, when he went back inside, he'd probably be stumbling about, half-blind, as you are when you come into a dark room after being outside in a bright light. He might even have difficulty interpreting the shadows he used to know so well. His companions, who had remained in the cave, would think he was a fool and would say that his visit to the upper world had ruined his sight and that the ascent was not worth even attempting. "And if anyone tried to release them to lead them up, they would kill him if they could lay hands on him."[2]

Ouch! So knowing more than the others wasn't exactly a benefit, was it?

I suppose it depends what you mean by a "benefit." Would you rather fit in with the group and miss out on the much bigger world outside? Think about the guy in H. G. Wells' story "The Country of the Blind." If he wants to stay alive in the beautiful hidden valley of the blind, he has to accept being blinded, so that he no longer talks like an idiot about the things he can see. In the end he flees the valley, but it was a tough choice, and he risked his life while doing it. But he felt that his sight was worth more than his security.

So he kept an extra sense that the blind hadn't got. And Plato's story gave the guy knowledge that the others in the cave hadn't got. But why are you telling us this in the context of different dimensions?

(We paused for thought for some time at this stage.) As the shadows on the wall of the cave are to a bigger reality of three dimensions so the visible reality of three dimensions is to . . . ? What is your answer? Think about it. Plato is saying that our understanding of reality may be as inadequate as the flat two-dimensional representation on a screen is compared to the three-dimensional world in which we live. And what physics is telling us is that we live in a world where we can only experience four dimensions, but there are actually ten. It's as if we are tied in a cave where we only see part of reality, a shadow of what's out there. Life has many more dimensions than we recognize, particularly if we insist on living in a world bounded by what we can see and touch.

Isn't it only recently that we began to say that only material things count as real? Hasn't religion always said there's more to life than meets the eye?

Yes, it has. It's actually quite interesting to look at the work of the great American psychologist William James who produced a famous critical evaluation of the belief systems of the world, called *The Varieties of Religious Experience.*[3] He found that every religion involved belief in an unseen order. In fact he summarized the central ideas of most religions: (1) The visible world is part of a more spiritual universe from which it draws its chief significance; (2) The purpose of living is to attain union or harmonious relation with that higher universe (which he calls God).

I suppose that is pretty much the same as Plato saying that what we ought to be doing is trying to find out more about the world that we can't see, and not being deceived by superficial appearances.

Right. And that's where the next three principles that James identified in all religions come in. First, they involve an uneasiness, a sense that there is something wrong about us, as we naturally stand; then comes an awareness that we can be saved from the wrongness by making proper connection with the higher powers. And the way this is done is through some form of prayer or inner communion with the Spirit as a process. He saw prayer as something where work is really done, and where spiritual energy flows and produces effects, psychological or material, within the physical world—which is pretty remarkable seeing that he himself was an agnostic.

What do you mean "an agnostic"?

He didn't accept any particular system of belief himself, but at the same time he found it impossible to deny that there are higher powers in the universe. He felt strongly that "materialism will always fail."[4]

His problem was how we could get any accurate knowledge of this outside dimension. Take the guy in the cave. He couldn't undo the shackles by himself, and if he had been chained up like that since he was a baby, he might not even have recognized that it was an inadequate way to live. He needed some outside help to get him out of the cave and to get him to survive long enough on the outside to be able to go back and tell the others about it. But what would that outside help be like?

Next time we meet, we're going to talk about the problem of communicating across dimensions. In the meanwhile, I want you to use your imagination a bit, and develop an allegory like Plato's. Suppose you and I, who are three-dimensional beings, had created a two-dimensional world we can call Flatland. How would we set about communicating with Flatlanders? What problems do you think we would run into? And how could we possibly solve them?

5. The Story of Flatland

Even I—who have been in Spaceland, and have had the privilege of understanding for twenty-four hours the meaning of "height"—even I cannot now comprehend it, nor realize it by the sense of sight or by any process of reason; I can but apprehend it by faith. —A. SQUARE[1]

WHEN THE GROUP arrived for our next session, I ushered them to a table scattered with paper squares, triangles, and pentagons, visual representations of the 2-D world we would enter to help us understand some of the problems posed by living in a world with unseen dimensions. And this time I was the one who asked the questions, while the students had to wrestle with the answers.

I told you that Flatland was a 2-D world, with length and breadth, but no height. What would it be like to live in a world like that?

I guess that there wouldn't be any "up" or "down" there. So there couldn't be a sky with a sun in it; there would just have to be light coming from one side or the other.

What do you think the Flatlanders, the people living there, would look like?

A straight line is one dimensional, so they might all just be straight lines. But all these paper shapes on the table are two dimensional, so they might come in different shapes and different sizes, but not in different heights. Maybe different colors too?

On second thought, no, to see color you would need to be above something to see its surface, and if this world is really 2-D, there is no "up" from which they can see each other. The most they can see is a little shadowing where their lines bend, depending on the direction of the light, but that would be all. Probably they would recognize each other by the length of a side and the number of angles they could see.

Most societies have some kind of social hierarchy, often based to some extent on looks. What do you think would be the most prestigious shape for a Flatlander?

I've heard that people everywhere tend to think that regular features make someone look attractive, and the most "regular" form in Flatland is a straight line.

But, on the other hand, if something is too regular, it's also called "plain." Curves are attractive too! I'd expect that the best shape to be would be a circle, which combines nice curves with regularity. It would also be the easiest shape to move around with: smooth rolling and no jarring corners.

So if I were to arrange these shapes in terms of prestige, I'd put the circles first, and then those shapes that come closest to circles. An octagon would outrank a hexagon, a square would outrank a triangle, and the poor lowly straight line would probably be right at the bottom of the pecking order.

When there is a social order, there is usually some means of moving up or down it? How might Flatlanders do this?

They'd have to increase their number of sides. People would try to get their kids to marry into the shape above them. So the triangles would want to marry squares, in the hopes that their children might at least be squares (or maybe even pentagons)!

But hopes for one's children aren't enough. If I were a Flatlander, I would want to get ahead myself. How could I do that?

A stretching and flexibility class might help you to put in a few kinks. But I doubt that exercise would be enough to give you angles. I've tried it in our 3-D world and am still no beauty! Plastic surgery might help me

. . . but I don't think I am prepared to endure the pain and run the risks of it not working. I bet the Flatlanders feel the same.

But if I could assure you that if I melted you down and recast you, you could be a circle, would you do it?
I am not sure. It might depend on how desperate I was. If I were a straight line, right at the bottom of the pile, with no hope of getting to the top, I might risk it. But if I were an octagon or decagon, who was "nearly there," I might not want to take the risk. Being broken up is risky, and I might be attached to a few of my angles.

Plus, if I did become a perfect circle, other people might be jealous or treat me differently (particularly if I'd just been at a much lower level before). On balance, I'd probably stick with what I know, and not take risks.

Now I am going to introduce a new element. Imagine that Flatland had been created by some being that was a sphere, a three-dimensional shape. The Sphere enjoyed creating that world and takes a real interest in what's going on there, but is bothered by the constant jockeying for status among Flatlanders, and the misery this causes. It would like to communicate with the Flatlanders as individuals and get them to fix this and other problems. How do you think it should go about it?
It would be really difficult because the Flatlanders haven't got any sense of anything that has more than two dimensions—it wouldn't make sense to them. To them, Flatland is the whole of reality.

I suppose it could just try speaking to them. After all, a sound has no dimensions; they could hear that.

But will they have any idea where the voice comes from?
They'll just think it's one of the rich neighbors with a megaphone! Or if they realize that's not where it comes from, they'll think they are going crazy, hearing things inside their head!

And if several of them heard the voice, but others didn't, they'd be accused of suffering from mass hallucinations or lying.

So why not avoid the problem by speaking in a way all the Flatlanders can hear at the same time?
The Sphere could do that if it wanted to. But that would be a bit like the principal speaking on the public address system. It's OK for delivering a message, but from what you said about the Sphere, I would think it wanted something more than that when it wanted to communicate with Flatlanders. Communication is a two-way thing, not just "I speak; you listen!"

What the Sphere needs to do is to find someone in Flatland who is sensitive enough to hear unexpected things and aware that there may be something outside Flatland, and then try to communicate with that person.

Suppose the Sphere had managed to find a few people like that. How do you think they would have reacted? What would others have made of them?
I think that the communication would have been a very important part of their lives. But I am not sure what they would have done with it.

Some of them might write down what they heard, or they might tell their friends what they were hearing from the Sphere, but their friends, if good friends, would probably just listen politely and ignore this part of the conversation. Alternatively, they might even poke fun at them, tease them about their voices, and generally give them a bad time until they stopped mentioning the Sphere in public.

So while the Sphere might be able to communicate with these one or two people, there wouldn't be any communication with most Flatlanders. Yet the Sphere wants to help fix some of the status problems that affect all of them. So the "voice" approach isn't very successful. What other options does the Sphere have?
Maybe the Sphere could consider taking one of the Flatlanders it has been speaking to, maybe a square, and somehow moving it into a 3-D world for a few minutes. The square would see the amazing complexity of the world, and would begin to understand something about the special relationship between Flatland and the Sphere. It would also be able to see patterns in the lives of its fellow Flatlanders that it hadn't

noticed before. When the square got back, it probably wouldn't be able to stop talking about what it had seen, and would try to encourage all the other Flatlanders to communicate with this marvelous Sphere too.

But wouldn't the other Flatlanders treat that square rather like the folk in Plato's cave treated the captive who returned? Would they believe anything he said?

A few might be impressed, and think about what he was saying, but most of them would probably just think the square was seeing things. Psychiatrists might get involved, and the poor square might end up in a mental hospital, to prevent him from doing any harm or "corrupting the minds of the young." (Hey, didn't they say something similar about Socrates?)

Even those who did consider believing this remarkable story because of the dramatic effect it had had on the square would die off in the end. The square's children might keep the memory alive for a while, but would it be passed on to the grandchildren, and the great-grandchildren? Would they accept it without any direct experience themselves?

So you think the vision would just fade away if only one person saw it. But what if the Sphere showed the three-dimensional universe to a whole lot of Flatlanders? Wouldn't that have a lot more impact?

Well, it would mean that the group could at least corroborate each other's stories when they got back. And they could talk to each other about the Sphere, and remind each other of what they had seen. So a small group would be more effective than an individual at getting the message out.

But look at what often happens to small groups who are excited about something in our world. The initial members are passionately excited about the cause, but those who come later are less committed—possibly because they haven't seen what the first group saw, but just have to take their word for it.

And the wider society tends to be suspicious of small, enthusiastic groups. It tends to isolate such a group, mock it, or even imprison its members.

Maybe the problem is that the Sphere is focusing on marginal groups, like squares and a few sensitive individuals. Maybe the Sphere should target a leader, someone other Flatlanders would respect and listen to?

It might work if the leader can really inspire loyalty—or obedience—in his courtiers and subjects. But leaders tend to fall from favor fairly easily, and the monuments they erect to their own ideas become museum pieces. What would happen if there was a palace revolution, and that ruler was overthrown? Would the long-term result really be much different from the effect of the Sphere revealing details to a few ordinary Flatlanders?

What about lifting all the Flatlanders up so that they could all see the 3-D universe that the Sphere lived in?

This might work, but if they were all taken to live in that 3-D world, Flatland would cease to exist. We might feel that the 3-D world would be so much better that they would never want to go back, but, then, Flatland wasn't our home. Plus the Sphere had created Flatland, and wasn't keen to destroy it.

If they only had a short visit to the 3-D world, how long would the effects last? Some of them would surely say, "Well, that was all very nice, but I have to live here in Flatland, and do what it takes to get ahead here." Any advice from the 3-D world about how to fix the problems in Flatland would probably be rejected by those who benefited from the problems.

Maybe the problem is that the Sphere isn't giving Flatlanders enough incentives? What if all the people who communicated with the Sphere became healthier, wealthier, and happier than everybody else? Surely the Sphere could organize that without too much trouble, and without really disrupting Flatland society?

But that leads to exactly the same type of problem you see in school. Everyone wants to be friends with the rich kid who dishes out party invitations and food—until the money runs out. It's very difficult to tell who likes you for yourself and who merely likes what you give if you are too lavish with favors.

Plus it sounds a bit like animal training, doesn't it. "Good dog—here's your treat." Is that the way the Sphere wants to communicate with people?

Surely, if the Sphere doesn't want mere hangers-on, all it has to do is flick a switch in the Flatlanders' minds to make them genuinely want to communicate? After all, the Sphere made them and programmed them. It obviously left a bug in the system that allows them not to want to communicate. So why not just fix that glitch?

That would be a quick and easy fix, wouldn't it? But it would also reduce them to the same status as robots. Is a friendship worth much if the other party has no choice about whether to be friends?

The Sphere's major problem is that it wants real communication with a group of beings who don't know it exists, and can barely begin to conceptualize its shape.

So it seems that the Sphere doesn't have many options when it comes to trying to communicate. It can either address all Flatlanders at once, which isn't a very personal approach to communication, or it can talk to just a few, but the message never gets out to a larger group. What should it do? Abandon the Flatlanders to their own devices? Destroy Flatland and create a universe where communication is easier?

You know, so far the Sphere has been trying to get people to listen, or trying to get them to look at the possibility of a 3-D universe. But maybe this is the wrong way to approach the problem. If the Sphere really wants to communicate, why not stop trying to get the Flatlanders to come to the Sphere's world and instead have the Sphere enter Flatland?

Would that be possible?

It wouldn't be easy. The Sphere would have to be prepared to sacrifice some of its important characteristics to do this. To go from a world with height and depth and breadth to a world that only has two of these is to lose a lot of beauty and wonder and possibilities.

And would the results be worth the risk? Sure, Flatlanders will be prepared to talk to you if you look like one of them, but will you be able to talk to them? Will they listen if you try to tell them about the other world that you know? What will it be like if they won't listen? In your own world, no one rejects you; on Flatland, somebody probably will. Is that the sort of experience you really want?

But those are all psychological issues. What about the physical issues? How would the Sphere get to be a Flatlander? Would it even be recognizable as a Sphere there?

The best way to work out what the Sphere would look like as it appeared in Flatland is to draw the Sphere in two dimensions. The Circle is the only part of the Sphere that the Flatlanders can see—a cross-section in two dimensions.

But even if all they can see is the circle, wouldn't the Sphere still have some 3-D attributes that Flatlanders couldn't see?

I suppose so, although the only time the Flatlanders would be aware of them would be when the Circle did something remarkable that involved the invisible third dimension, for example, rearranging reality by changing angles or repairing broken shapes.

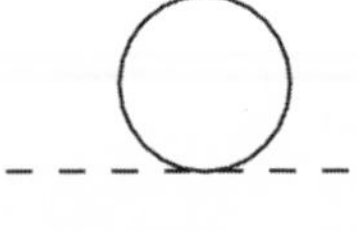

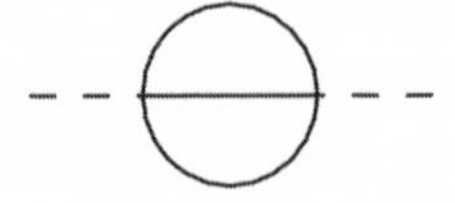

FIGURE 5.1

So the Sphere has now established a personal foothold on Flatland, and is in a position to communicate with Flatlanders. What would be the content of that communication?

Certainly, he'd tell them that there are more than two dimensions, and he'd also tell them about the existence of an invisible but creating Sphere who cares deeply about Flatland and wants to communicate with its inhabitants and change their lives for the better.

If what had been bothering him before was the hierarchical jockeying for social status with everyone wanting to be circles and putting down those in the lower shapes, he might undercut this by pointing out that the perfect shape wasn't a circle but a sphere. And that no matter how hard they worked at getting to be circles, spherehood was beyond them without his help.

How do you think Flatlanders will react to this message?

Some of them would be attracted to him. After all, he was probably the closest to a perfect circle they had ever seen, and the remarkable things he did and taught would have caught their attention.

Plus, if you were one of the groups who had no hope of ever getting to be a circle, it would be great to hear that becoming a circle wasn't the be-all and end-all of life. If he also offered to help you become a little sphere, even if you were just a triangle before, it would be pretty exciting stuff.

Some of the polygons might also have been interested in discussing the notion of the existence of a Sphere with this new Flatlander. Some of them might have had ancestors who had reportedly heard the Sphere, and they would have been interested in the science and philosophy of what he had to say.

But a lot of the others would not have liked him at all. What he was teaching was going to upset the whole social hierarchy. The polygons had enjoyed leadership for years, and now lowly triangles were suddenly talking about developing curves! Nor would they be impressed that all their efforts to get more angles and closer to a circle in shape were being dismissed as nowhere near good enough if the perfect shape was a sphere.

There would also have been plenty of Flatlanders who would argue that the things the Circle was talking about would destabilize society and disrupt the economy. For one thing, all those manufacturers of exercise equipment and training manuals would be put out of business if the Circle started transforming people into spheres, whatever those were.

And it wouldn't be just the octagons who would feel this way: everyone who was higher than a mere line would lose some status if everyone became circular. And they would lose the world they knew and were more or less comfortable with most of the time.

So what do you think they will do about it?

Try to discredit the Circle and limit the harm being done to their position.

The polygons would probably try to avoid violence if at all possible, but they would remind themselves that the peace and security of the whole are more important than the rights of one individual.

How do you think the Circle (who we know is a Sphere) will react to this threat?

He could easily simply put on his "spherical" boots in that third dimension and crush those polygons like beetles, couldn't he? But what would the result of that be in the long term? Some Flatlanders would be impressed, briefly, but others would be disturbed by his violence. And chaos and war would undoubtedly break out. Was that what he had come to accomplish? Violence tends to work against communication.

No, the Circle would decide not to use power to retaliate. After all, this rejection was not completely unexpected, although no more pleasant for that! The powers-that-be would be permitted to work to destroy him.

But would it really be possible to destroy something that exists on three dimensions in a two-dimensional world?

Probably not, though you could probably do serious damage to the slice, the circle, which is all that the Flatlanders can see of the Sphere. The Circle would be hurt and shamed: physically and mentally. But the spirit of the Sphere would be intact.

Now what! What does the Sphere do? The polygons think they have destroyed it. Will it decide that enough is enough and go back to the three-dimensional universe and forget about the Flatlanders, or decide simply to wipe them out? If you were the Sphere and they had rejected you, what would you feel and what might you do next?

If it were I, I wouldn't have let them hurt any part of me in the first place! I would have wiped them out before they even got close. But the Sphere didn't do that then, and so probably won't do it now either.

I also wouldn't have put myself at risk by caring about the Flatlanders either. But the Sphere seems to be really interested in helping them, and getting involved with them. And having lived there for some time, it will know them in a different way than it did before, and will be even more anxious to reach out to them.

I think the Sphere will have to find some way to prove to them that they really haven't killed it—that I had met the challenge of the reactionary forces and is bigger than death itself. Even if the polygons don't believe this, the Sphere's followers will, and will realize that what the Sphere said about this life being only part of a much bigger reality is perfectly true.

OK. So the Sphere isn't going to give up. So what will happen next?

If I were the Sphere, I'd come back a few times to prove that killing the Circle hadn't killed me. But I suppose when I came back I would have to look like that Circle again, so that they could recognize me.

I think I would also like to rub in the lesson that I was alive and bigger than they were—not in a hurtful way, but in way they'd recognize. The best way to do this might be for the Circle to show up unexpectedly, in places where he used to hang out. But I wouldn't merely dodge around in the background and then appear, as if I were a fugitive. I'd simply use my extra-dimensional powers to just suddenly be there. That would prove that I really existed, and that there was a bigger reality than they had counted on when they killed me, but it wouldn't hurt anyone.

But you know, even if I didn't want to hurt those who had opposed me, I might want to reward those who had believed what I told them, and who had been so bitterly disappointed when they thought I was dead. The best way to do that might be to take those Flatlanders and give them an opportunity to see the Circle in the higher dimension and to experience the power he has there. Can you imagine how excited they would be, and how much faith they would have in his promise that he could enable them to live in this dimension too?

They'd love it! They'd want to stay there permanently. But would the Sphere allow that?

I don't think the Sphere would want these believing Flatlanders to leave Flatland completely. After all, he had gone to all that trouble just to make contact with them, so why would he remove his live contacts just when the communication was really starting to flow?

So he'd send them back from his dimension, but their new understanding that he had created Flatland would mean that their relationship with him would be quite different. He had told them that his ultimate goal was to make them into miniature spheres, and to do this he would need to knock off corners and change their thinking about "the way the world works." He might even want to change the material from which they were made into something more malleable. Once they had seen his 3-D reality, they would have much more trust in what he was setting out to do to them.

Also, because they had seen him appear unexpectedly, they would know that he could carry on doing this, and could appear unexpectedly anywhere in Flatland. He could carry on working through later generations of Flatlanders, working to turn them into miniature spheres too. Even if they couldn't see him, they would know that his 3-D world was very close to theirs, and that they could talk to him directly.

Does any of this sound familiar?

Yes, it does sound vaguely like what I have heard Christians talk about. God making the world, and trying to communicate with it through guys like Moses and the prophets, and then Jesus. I suppose this makes a bit more sense of what Christians mean when they say Jesus is God come to earth. So I suppose you are saying that if Jesus is like the Sphere/Circle, who comes back after he is killed, then we are like the Flatlanders, a bit blind to a bigger reality that Jesus is trying to tell us about?

Yes, you're on the right track. Let's try another example. Do any of you know the story of Saul, who became Paul? He was a guy who was trying hard to be perfect. He was born into a good family, kept the rules, and was near the top of the social scale. How would you fit that into the Flatland analogy?

I guess Saul is like one of those octagons, maybe even close to a decagon. Because he was so close to being a circle, the one thing that would really have annoyed him would have been a bunch of Flatlanders who kept announcing that aiming to be perfectly circular was not enough, but that you should become a miniature sphere. And that the Circle would help you get there, regardless of your present shape.

And what happened to him?

Well, one day this octagon was heading off to cause trouble for some Circuits (I suppose that's what Flatlanders who believed in a Circle might come to be called) when suddenly the Circle/Sphere himself appeared in front of Saul. At one and the same time Saul recognized the Circle for who he was and became blind to the Flatland he lived in.

Saul's Flatland sight eventually came back, but he never lost his view of the world of the Circle/Sphere either. Nor did he ever forget that he couldn't get to be a miniature sphere by his own efforts: he needed the Circle's help. He changed his name to Paul and spent the rest of his life creating what I suppose we could call new circuit boards—new churches where others could encounter the Circle too.

Well done! What about going home and trying to see whether the Flatland analogy fits with other parts of the story. Maybe you could try reading one of the Gospels, and working out the pattern, in the same way as we did for Saul. If you find it's really difficult, ask the Spirit to help you.

I suppose we could do that. It's a great story that would make perfect sense—if only it were true. But Plato's metaphor of the cave was just that, a metaphor. And all that we have done here is make up another story. I'll grant you the fact that there may be extra dimensions we don't know about. The physics has proved that, and Flatland does show how difficult it would be to communicate across dimensions. But we still have to assume so much. I mean, we have to assume that there is a Creator who cares about the world and wants to communicate with humans. That's a huge statement of faith.

You're right, although the leap of faith isn't quite as huge as you suppose. After all, we do have four different eyewitness accounts of Jesus' life, which you can compare when investigating whether the Flatland metaphor fits. But the second part of your question needs another whole session to discuss!

The students' question about whether there was any evidence not just for a Creator but for a caring Creator dictated the topic for our next session. We reluctantly abandoned Flatland,[2] and turned our attention to the anthropic principle—the science that suggests that maybe there is something special about life on this planet.

6. The Anthropic Principle

Some supernatural intelligence must be directing the evolution of life and indeed the whole cosmos. The universe is an obvious fix—there are too many things that look accidental that are not. —SIR FRED HOYLE[1]

OUR DISCUSSION OF Flatland had raised the question of whether it was reasonable to think that people might have a special place in the universe, or be a particular concern of a Creator. So in this session we set out to explore whether it was more logical to adopt the materialistic view that human life is a total accident, or whether there is any evidence to support the view that the universe might have been designed to support human life.

Don't you think it is really arrogant to assume that the universe was made for us?

Yes, of course it is. But merely slapping the label arrogant onto an idea doesn't mean you don't need to consider the evidence.

Over the last twenty years or so, scientists have become aware of the astonishing fact that the laws of physics seem to be extremely fine-tuned to allow life here on earth to develop. If there were only slight changes in the values of any of the fundamental constants, aspects of the physical world, we wouldn't be here. So now it's not just religious people who talk about the world having been made for humanity; serious scientists are also debating the same thing, but calling it the anthropic principle.[2]

So are you saying that everything since the big bang has focused on producing us?

More than that. I am saying that everything including the big bang has been directed to producing not just us but all the life on earth.

We started this book by talking about the big bang; now let's look at it again. If the universe at, say, one second after the big bang had been smaller by even one part in a thousand billion, it would have collapsed before it ever reached its present size. If it had expanded faster by even the same tiny amount, matter would have been so thinly dispersed in space that no galaxies or stars could have condensed. Neither supernovae nor planets could have formed, and the universe would eventually have collapsed again. If the expansion rate of the universe just after the big bang hadn't been so finely tuned, there wouldn't be a universe.

Many scientists are impressed by the accuracy of the mathematical equations—accurate to one part in many billions of billions. As Roger Penrose comments, "The Creator, if there was one, must have had a pretty good idea of how he wanted the universe to begin."[3]

What about those ten different dimensions at the time of the big bang? Did they also balance out like that?

Physicists are confident that, in the very earliest evolution of the universe, at the initial singularity, there were ten (or even eleven) dimensions, not just four. These ten were all equal, and they could have divided in two possible ways. Three of them could have become compacted while the other seven expanded (to give us six space dimensions plus one time dimension). Or six of them could have compacted and curled up at each point in space, while the remaining four expanded to give us three space dimensions plus one time dimension (which sounds a bit more familiar).

It has been suggested that stable atoms, chemistry, and life as we know it can only exist in a space with three dimensions.[4] We cannot conceive of a world with six dimensions (except in terms of a "new heaven and a new earth").

But just because we can't conceive it doesn't mean it can't exist. After all, the world could have been made of antimatter rather than matter, and we can't imagine that either.

You're right. But there is another amazing balance there too. One would have expected that at the big bang the amounts of matter and antimatter produced would have been almost the same. Of course, the problem is that if they were present in equal amounts, they would immediately have annihilated one another in the kind of cataclysmic explosion of energy that helped to fuel the big bang itself. Nothing would have remained afterward. We can see proof of this on a small scale in particle accelerators and cyclotrons, where protons and antiprotons, electrons and positrons, quarks and antiquarks destroy each other. The universe would have died at birth.

However, the laws of physics have a very slight asymmetry built in from the beginning. For every billion particles of antimatter created in the big bang, a billion and one particles of matter were created:

10^9 of antimatter = $(10^9 + 1)$ of matter

This tiny imbalance giving us one extra particle of matter is responsible for all the matter in the whole universe. The leftovers of the catastrophic explosion caused by the mixing of matter and antimatter slowly coalesced into stars and galaxies. If the asymmetry had been any different, the universe would not have produced the present conditions—and us.

But isn't this all just a coincidence. Without the big bang, the universe wouldn't have been here at all. And if the big bang had blown differently, there would be a different universe. So I still don't see that proves any anthropic principle. Surely if the universe was created for us, it would have produced us a bit faster? The universe is millions of years older than we are.

Good point. But if you think about the physics, a long "cooking time" was essential. It takes about 10,000 million years for first-generation stars to evolve and produce galaxies and supernovae. The debris from those explosions then had to reform to produce the newer stars and planets, like those in our solar system, which is over 4,000 million years old. Our solar system needed our own galaxy to allow for that first generation of stars to form supernovae that created many of the elements we need for life. We needed all the other galaxies because the initial rate of expansion must be very, very precise to avoid the universe collapsing in on itself again.

We also can't take it for granted that life would automatically follow from these supernova explosions, for the burst of neutrinos that blew apart the outer layers of the star had to be precisely the right strength to disperse the essential ingredients of life across the galaxy. Other circumstances also had to be exactly right: as the *New Scientist* reported, "The window of opportunity for a universe in which there is some helium and there are also exploding supernovae is very narrow!"[5]

What do you mean when you talk about supernovae creating the essential ingredients for life?

All of the elements essential for life, except hydrogen and a little helium, were manufactured inside stars, and spread across the universe when those stars exploded. Your body is composed of stardust!

We now have a pretty good understanding of how stellar nucleosynthesis works, and there are some amazing coincidences there too. For example, the existence of carbon.

You will have heard people say that we are a carbon-based life-form, and they are right, but there could easily have been far too little carbon in the universe for any carbon-based life to develop. For years, scientists struggled to work out how the helium and beryllium that formed inside stars could possibly be converted to carbon. Working with what is called nuclear resonance, Fred Hoyle determined that the thermal energy in the core of a star is precisely what is needed to force these elements to combine to create carbon. Had the energy been slightly lower, no carbon could have been created; had it been slightly higher, we would have had abundant oxygen instead of carbon. Without abundant carbon, we could never have the marvelous flexibility of the DNA molecule. Its creation was an astronomical hole-in-one!

Are there other chemical coincidences that seem to be designed to encourage life?

Well, there is the fact that life-making materials seem to be available just where we need them. Without an abundant supply of carbon, nitrogen, phosphorus, and oxygen, there would be no self-reproducing molecules of DNA (deoxyribonucleic acid) to control the rate of synthesis of complex organic molecules. The stars supplied those raw materials. And that coincidence is linked to an enormous number of other interlocking "coincidences" and conditions that led to the formation of DNA at all.

Then there are all the heavier elements our bodies need: iron, molybdenum, zinc, cobalt, besides sodium, magnesium, potassium, and the like. They were all here where they were needed.

And don't forget that amazing stuff, water. You all learned about the life-giving water cycle in primary school, but do you realize how finely poised the properties of water are? The hydrogen bonding keeps it liquid, but the evaporation and freezing points are precisely where we need them to be if we are to survive.

Oh yes, and when talking about the chemistry that is needed to make life, don't forget the chemicals we don't have. Here on earth, we don't have the clouds of poisonous methane or ammonia gases that are found on many other planets that would prevent life.

That is quite a string of coincidences. Are there any nonchemical coincidences too?

Plenty of them! The ratio of matter to radiation is one. So is the fact that even slight differences in the strength of the four fundamental forces of nature (gravity, electromagnetism, the strong force in the nucleus, and the weak force of radioactivity) would make chemistry impossible—there would be no stars, no carbon compounds, and therefore no observers.

Most of what you have talked about so far relates to life in the universe. What about life on earth? Are there any things about the earth that suggest it might have been designed for life?

Again, there are a whole string of them, many of them related to energy. Life needs a continuous supply of free energy. This energy has to be stable over a long period, with no fluctuations—it's fortunate that the sun burns with extraordinary uniformity and that the earth's orbit is nearly circular, so that the supply of energy is fairly constant.

Human life can also only exist within a fairly narrow range of temperatures, roughly between 5°C and 40°C degrees (unless we wear special clothing). Fortunately for us, our distance from the sun and the angle of the earth's tilt are exactly right to produce temperatures in this range, and to give us day and night and different seasons.

So we have a supply of energy from space, but we also have some helpful energy from the earth itself. The earth's gravity is strong enough to stop our atmosphere from evaporating into space, but weak enough for us to move about easily.

And this gravity also keeps the ozone layer in place above the atmosphere, where it blocks the deadly ultraviolet radiation from the sun, so that we get just the amount of sun energy we need. (Of course, recently humanity has been damaging this ozone layer, with disastrous effects.)

It's not just the ozone layer that protects us either. Without the earth's protective magnetic field, harmful cosmic subatomic particles would deluge the earth's surface and kill us off. In fact, astronomers have recently gained a new appreciation of how deadly our galaxy can be—and how fortunate the earth is occupy a narrow habitable zone where it is relatively safe from being hit by asteroids, comets, and blasts of radiation. The massive planet Jupiter and our large moon are powerful defenses. And we should also be very thankful that the supermassive black hole at the heart of our galaxy is currently relatively dormant.

In fact, the more we look at things, the more we realize how rare the conditions that make for intelligent life must be. While it is possible that microbiological life (at the level of bacteria) is common, even widespread throughout the billions of stars and galaxies, there are likely to be very few, and maybe no, other planets with such a rich biosphere and evolving intelligent life.[6]

It's certainly quite an impressive chain of coincidences—enough to make one start thinking about the anthropic principle. What do scientists make of this?

All of them are in agreement that "if it had been different, we would not be here."[7] Some, however, would hold to the weak version of the principle and others to the strong.

The weak anthropic principle states that our own existence requires certain necessary conditions to be met regarding the past and present structure of the visible universe. These numerous "coincidences" are essential for the existence of observers.

Brandon Carter, who originally formulated the weak principle, later went even further in formulating the "strong" anthropic principle, which states that the laws of nature and the cosmic arrangement are constrained in such a way that intelligent observers must inevitably come in to being.[8]

So what conclusions do you draw from all this?

It seems that there are a couple of possible conclusions. Some scientists, like Stephen Hawking, would suggest that it is merely an incredible stroke of random luck that life emerged at all.

Others have suggested that ours is just one among many parallel worlds, most of which would not have any living beings. As we saw in the chapter on quantum physics, Wheeler originally supported this many-universes theory, but has abandoned it, as have many others. Even an avowed agnostic such as Hermann Bondi[9] pours scorn on it as a completely untestable theory.

Still others have said that the incredible chain of coincidences required to produce life can only be interpreted as evidence for some sort of design in creation. Even Sir Fred Hoyle, formerly an avowed atheist, now affirms that "a common-sense view of the facts suggests that a superintellect has monkeyed with physics, as well as chemistry and biology, and that there are no blind forces worth speaking about in nature."[10] Even more, he famously insists that the spontaneous generation of life on earth is about as likely as the "assemblage of a 747 aircraft by a tornado passing through a junkyard."[11]

Similarly, Paul Davies comments that the fact that the universe is fine-tuned to such stunning accuracy is "surely one of the great mysteries of cosmology . . . [I]t is hard to resist the impression of something—some influence capable of transcending spacetime . . . which possesses an overview of the entire cosmos . . . quite beyond rational explanations based on physical theory."[12] Design is the best explanation for all cosmic fine-tuning. Judicious design is also confirmed by Paul Davies in his latest book, *Cosmic Jackpot: Why Our Universe Is Just Right for Life.* The only way out seems to propose billions and billions of universes—the multiverse theory—with an infinity of universes, and an infinity of almost-earths. This seems a very contrived explanation, little different from science fiction.

While there is no final proof there is a Creator, the possibility that one exists seems less and less implausible. And if there is, that Creator has gone to an enormous amount of trouble to create a world in which life can exist. It does then seem plausible that, as in the Flatland analogy, the Creator might take an active interest in what that life gets up to.

It really seems that the argument for God is a lot stronger than I ever thought. What you have said so far does make the mere existence of life seem almost miraculous. But it still doesn't answer questions about whether human life, our lives, are at all important. Surely evolution says that we are just another leaf on the tree of life, and that all life is equally special—and amazing.

Good point. And that is why we are going to be talking specifically about evolution in our next session. But there is also one other quotation I want you to mull over in the meantime. It's from the Astronomer Royal Martin Rees, who notes that "unlike all previous generations, we know how we came to be here. But like all previous generations, we still do not know why."[13]

7. Evolution: Where Do We Fit In?

Humans are distinguished from the animals only in their spiritual qualities. Consequently it is reasonable to suppose (and Genesis 2 suggests it) that God placed His image in an already existing animal. —R. J. BERRY[1]

WE HAD TALKED about the amazing precision required for our planet to emerge as a place where life could develop. But we hadn't discussed how life actually came into being, or whether human life is in any way different from animal and plant life. These were the questions the students wanted to discuss next.

We have heard a lot about physics and chemistry so far, but what about biology? What do we know about how this planet moved from being merely a collection of chemicals with the potential to form life to actually producing life?

Not a lot. We don't have any clear answers yet. All that we do know is that the earth was formed about 4,500 million years ago from debris that circled the sun. It took some time for this material to condense and cool, for rocks to solidify, for the waters to separate into oceans and thick cloud, while the land began to cool and contract as mountains and volcanoes emerged from the seas.

Somewhere in this volcanic wasteland, the first miracle of a progressive journey to life occurred between 4.2 and 3.9 billion years ago as the first living cells appeared. There is no life without cells, but we have no idea how the first cell came to form. We do know that the first cells were probably bacteria living in deep undersea thermal vents or deep underground—even within solid rock itself. These bacteria were able to work without sunlight, and they began to feed on the metal-rich rock, turning hydrogen and sulphides into the poisonous gases methane and hydrogen sulphide.

Gradually, other photosynthetic bacteria developed, which were able to use the sun's energy to turn the carbon dioxide in the atmosphere into oxygen. Then began the slow business of filling the atmosphere with oxygen. Once oxygen had formed, an ozone layer also formed, which shielded the earth from lethal radiation, enabling a different kind of life to begin.

Incidentally, don't think that the role of bacteria in making the world habitable ended eons ago. We are still very dependent on the bacteria that fix nitrogen from the air and make it available to plants and on the tiny arthropods and fungi that dismantle the dead to form the nutritious dust for the next round of life. Bacteria are true living fossils, and also the most abundant living forms on earth (there are as many bacteria in a drop of water as there are humans alive today).

How did we get from single-cell bacteria to plants and animals?

How we did it, we don't know. But what we do know is that it involved the emergence of eukaryotic cells, that is, cells that have a nucleus with the wonderful genetic material, DNA, and special structures called mitochondria that produce energy. This was an amazingly difficult step. It involved one organism learning to live inside another. The simple cell membrane of the bacteria had to develop the ability to engulf other large organelles like mitochondria and ribosomes, store them within a membrane, and keep them alive by developing anaerobic respiration. No wonder the process took so long! These eukaryotic cells gradually formed multicellular plants and fish in the shallow water habitats produced by the new land surfaces being formed by continental drift about two and a half billion years ago. In another of those amazing "coincidences," the tides produced by the moon created a relatively stable intertidal zone along the shore where life could become established. Plants developed the important ability to turn the carbon dioxide in the atmosphere into fibers and sugar, which would eventually provide food for animals.

This process must have been incredibly slow! Wasn't there some sort of "big bang" for life, as there was for the start of the universe?

It was very slow at first. The first bacteria probably appeared about four billion years ago, and the multicellular organisms probably developed some 600 million years ago. But it is very difficult to find any fossil records of these organisms. But then in the rocks from the Cambrian period about 530 million years ago, we suddenly find the fossilized remains of a vast army of animals, plants, and life-forms. Some of the primitive mosses and ferns that grew in that era are still thriving today.

And from then on evolution just carried on grinding till we got the apes and then us, right?

Only partly. There is a link between us and other animals. Our DNA proves it—it's 98 percent the same as that of bonobos. But what a difference that 2 percent makes! And the history of the human race is a lot more complicated than your straight line of evolution supposes.

So how did we get to have humans around?

Answering that question is like being asked to summarize a book that has had most of the pages torn out and is written in code that we are not always sure how to translate! I can give you some information, but there are huge areas about which we know nothing.

Let's start off by talking about hominids, all those species that look sort-of human. Over the past 4 million years, perhaps around twenty hominid species have emerged, competed, coexisted, moved to new environments, and succeeded for a while. The earliest fossil remains we have come from Ethiopia and date back over 3 million years. We have an almost complete skeleton of a 3.3 million-year-old hominid, *Australopithecus afarensis,* found in Ethiopia, a three-year-old girl nicknamed "Salem." This is the same hominid species as the famous "Lucy" skeleton discovered in the same region in 1974. The legs and pelvis of Salem indicate the ability to walk upright on two legs. The upper part of the body is partly arboreal and the voice box shows that speech had not yet started to evolve. Most researchers believe that bipedalism is indicative of early members of our own genus, *Homo sapiens.* Salem gives insight into human evolution, where different parts of the hominid body were undergoing adaptation and selection at different times. Arms, shoulders, and speech developed at a later stage.[2] These very early hominids were small brained with large cheek teeth and seem to have died out about two million years ago.

Another species that seems to have overlapped with *Australopithecus* was *Homo habilis*, discovered by Louis and Mary Leakey in Kenya in the 1960s. This creature, which lived some 2.8 million years ago, had a larger brain than *Australopithecus* and small cheek teeth, and may have used primitive tools (which was why it was given the name *habilis*). About two million years ago, at the same time that *Australopithecus* and *Homo habilis* were disappearing, we begin to find fossils of more humanlike species with still larger brains who seem to have used slightly better tools. First we find *Homo ergaster*, and later *Homo erectus*, or "upright man," who seems to have walked in the same way we do, but who died out about one million years ago. Fossils of these hominids have been found in Africa and in eastern Asia, which initially made anthropologists think that these human-looking creatures had spread all over Asia from Africa, and then had evolved into modern humans. But this theory, which was standard in the late 1970s, has had to be abandoned because of the discovery of the African Eve.

I've heard about the African Eve. Is she the same as one in the Bible?

In some ways, yes. She is named after the Eve in the Bible because, like her, she is "the mother of all humans," or at least, that is what the science indicates. In the 1980s Allan Wilson and his colleagues at the University of California at Berkeley made a detailed study of human mitochondrial DNA, inherited through the mother, which indicates who is related to whom. To their surprise, their results led them to conclude that we are all descended from a single woman who lived between one and two hundred thousand years ago, almost certainly in Africa. She may have been a member of a group of modern humans, *Homo sapiens*, characterized by walking upright on two feet and by having a larger and more complex brain. We know that this group was developing in Africa around 100,000 to 140,000 years ago.

Incidentally, if you guys are worried about your place in history, you may be reassured to know that new studies in the variation of the Y-chromosomes, which reflects the male inheritance, also place "Adam" there at the same time![3]

Sometimes I'll describe someone I don't like as being "Neanderthal." Where does that term come from?

The Neanderthals were a separate branch of the hominid tree, who emerged around 200,000 years ago and coexisted with *Homo sapiens*, our branch, until about 30,000 years ago. They were the closest branch to us, and they seem to have lived in small communities, decorated their bodies with red ochre, and occasionally buried their dead. This last act is probably the most significant, because it proves they had self-consciousness and the ability to deal with abstractions. But analysis of the DNA in the fossil bones of Neanderthals shows that they were not identical to modern humans. In fact, the 2 percent by which they differ from us is the same as the amount by which we differ from chimpanzees. At most, we share a common ancestor in *Homo heidelbergensis*, an offshoot, with *Homo erectus*, of an earlier hominid (perhaps *Homo habilis*). The Neanderthals eventually died out about 30,000 years ago.

In October 2004, scientists announced the fascinating discovery of a three-foot-tall female *Homo floresiensis* on the island of Flores in Indonesia. This new human species, nicknamed "hobbits," was smaller than its ancestor *Homo erectus* and recently confirmed as a separate species.[4]

So that left just us.

Yes. *Homo sapiens* is the only hominid species to have survived, and we probably came out of Africa no more than 100,000 years ago. And for a long time, there wasn't that much remarkably different about us—we were just another set of hominids wandering around the world. And then about 50,000 years ago we start to find evidence of radical changes in tool building and signs of communal living and perhaps of language. This was followed in the Middle/Upper Paleolithic period (what you probably call the Stone Age) by a cultural explosion, a revolution in innovation. Between 35,000 and 15,000 years ago we suddenly find wonderful cave paintings, carvings, primitive whistles and flutes, more elaborate burials of the dead, and other signs that indicate an increased awareness and flowering of the human intellect. We were changing from merely clever hominid (humanlike) toolmakers to potential masters of the planet.

It seems you're saying that toolmaking isn't the characteristic that distinguishes us from animals, but that art and care for the dead are. What about language?

Language is a fairly recent development. It seems unlikely that even the Neanderthals, our closest relatives, had much ability to use language. But in *Homo sapiens* the larynx shifted to a unique position, which gives humans the distinction of being the only mammals incapable of drinking and breathing at the same time, and increases the chance of their choking to death while eating. But in exchange, humans received the ability to articulate a much wider range of sounds, both vowels and consonants, to develop language. Once the possibility of language arose, it would have spread rapidly by social contact.

If humans really can't drink and breathe at the same time, how on earth do babies survive when they are sucking on the breast or the bottle for a long time?

Fortunately, in newborn babies the larynx is still in a higher position, so they don't choke easily. It "drops" after six to eight months, which is when babies start to become able to produce more language-like sounds. This may well reflect the development of language in our forbears! So may the fact that children seem to develop language spontaneously in response to their environment. At first this is a fairly primitive protolanguage, which rapidly develops into the structured language that distinguishes us from all other species.

And, of course, there is another aspect of language that seems to distinguish us from other species: we seem to be the only ones who can internalize speech and so create an inner voice—what you might call a conscience, or consciousness. The complexity of human self-awareness became our springboard to true humanity.

But why did this happen only in humans? Why haven't other animals carried on developing in similar ways? We talked about the anthropic principle in cosmology and biology. Is there something similar in relation to humanity?

We don't fully understand the levels of consciousness of other animals, but it is clear that they don't engage in abstract thinking or in creative acts to anything like the same extent that humans do. Why not? We don't know. Some force moved hominids in this direction.

The strong anthropic principle, which we discussed before, states that the laws of nature and the cosmic arrangement are constrained in such a way that intelligent observers must inevitably come into being. It certainly seems as if we must have been pushed in this direction, for the odds against our having ever appeared spontaneously are quite overwhelming. Design seems to have been at work from the instant of the big bang until humans emerged with the capacity for love, appreciation of beauty, and consciousness.

Keith Ward of Oxford University speaks of the unconscious purposiveness in natural processes, such that a process that seems largely random in fact provides a finely balanced environment for the development of conscious life-forms. He argues that the evolution of consciousness would become highly probable if selection within the possible mutations available at any time "was influenced by an intelligent design."[5]

You are talking about evolution as part of design, but I have heard that some Christians are totally opposed to the very idea of evolution. You don't seem to have a problem with it.

I think most people, even the most fundamentalist Christians, accept that some evolution has definitely taken place. After all, it's still happening around us—bacteria such as MRSA are able to evolve and multiply by a new highly virulent strain that becomes resistant to existing drugs.

Another striking example of evolution in action is the small lizard known as *Anolis sagrei*. When researchers introduced a larger, predatory curly-tailed lizard to the Caribbean islands where they live, *A. sagrei* initially began to evolve longer legs for speedier escapes. But then they began to seek refuge in trees, and shorter legs were found to become the favored trait to help climb more easily.

But that's only a very simple example of a change within a species, and doesn't involve any deep genetic change of the kind that produced humans. What about the type of evolution in which one species develops into another, and hominids become* Homo sapiens? *I thought Christians objected to that?

Some do, but sometimes what they are objecting to is the way the debate is framed, rather than the process itself. After all, even Darwin himself saw no good reason why his views should shock anyone's religious feelings: "There is grandeur in this view of life . . . having been originally breathed by the Creator into a few forms or into one."[6]

Many are happier with the word "development" or "development from previous material." Others assert that there are no macroevolutionary jumps. All processes are microevolutionary to produce different phyla through intermediate designs. There is a current emphasis on "evolutionary convergence"—the direction of evolution, converging to the same solution, is very striking at many levels. The course of evolution is not as random as some might think. Some paleontologists have called the phenomena "eerie" or "uncanny."[7] It has direction to it, rather than a purely random process. Materials at one time useful for one function are then co-opted or redeployed for a different function. None of this proves the existence of God, "but it all fits together." "Trends in evolutionary biology imply directionality and perhaps progress."[8]

However, some biologists, like Richard Dawkins with his "selfish gene" theory, insist that evolution is an entirely mechanical process, driven by a set of coincidences.[9]

There is no Creator, and thus no design. The concepts of ethics and morality that distinguish humans from animals are merely a product of the drive for self-preservation, which has discovered that some altruism is good because it enhances the chances of survival. One problem with this approach is that its supporters aren't completely consistent: Dawkins cares passionately about truth, beauty, and elegance in the laws of nature, about goodness, intellectual honesty in facing unpalatable facts—but there is no way he can derive these values from the purely materialistic worldview to which he claims to be committed.

My own problem with this view is that it's difficult to see how a mindless, purposeless, mechanical process of natural selection can really account for the superb design of living things, for music, art, and love. This view also doesn't face up to the sheer improbability of our existence. How can a simple primeval system reach such a high level of organization without the help of a designer or preexisting mind?

Another view that Christians also object to is the idea that God is merely the transcendent "watchmaker" who wound the world up, calibrated the big bang, and then left nature entirely to its own devices.

Do you remember the Flatlander model? If you had been the Sphere and had gone to the trouble of creating a 2-D world, wouldn't you have been interested in seeing what was going on there, and nudging it to make it more interesting, yet seeing the uncanny ability of evolution to reach appropriate solutions through all the possibilities?

Many Christians have no problems with the broad picture of evolution, because they recognize the need for both creation theory and a mechanism for development. The emergence in complexity from the very first pre-life beginnings defies full explanation. Physical answers "seem appropriate at one level of explanation only."[10] They are not alone in this view, for many scientists share it, for example, Steven J. Gould, who argues that strict Darwinians have overemphasized the role of natural selection and ignored other options, such as random genetic drifts and wayward meteorites. He cites the example of the Permian extinction 250 million years ago, which may have been caused by an exploding supernova and wiped out 90 percent of all species on the earth. Another global convulsion about 65 million years ago, probably caused by an asteroid blasting into the earth, wiped out the dinosaurs and introduced the age of mammals.[11] The resultant warming brought the dawn. R. J. Berry, professor of genetics at University College London, agrees with Gould, although he also sees the mystery of divine creation through natural processes.[12] Roger Penrose, too, feels there is something mysterious about natural selection because living things evolve "better than they ought to."[13]

But isn't what is called "punctuated evolution," with God doing the punctuation, really just a "God of the gaps" approach? If you don't understand something, you just assume a miracle occurred. But people used to think that an awful lot of things were miraculous that we can now explain quite easily in terms of science.

True. This is a mistake that people who believe in God have fallen into in the past, and will probably repeat in the future. What's important is not to get hung up on particular points, but to look at the long chain of improbable coincidences that got us to where we are now. We can't always say whether the chain was set up from the beginning so it would get us here, or whether nudges were needed along the way. But the mere fact that anything at all is alive, and that we ourselves have moved so far from the bacteria that we have self-consciousness, the ability to think in abstract terms, the capacity to love, and the capacity to create beauty—or ugliness—all of these are more easily explained by design than by random tides that threw us up on the shores of life. Humans, as well as matter and animal life, "can legitimately be regarded as special creations."[14]

So are you saying that the Creator started things off at the big bang, and then every now and again he may have reached into the world and tweaked things to send our development in the direction he wanted?

Perhaps at key points such as "Adam," and, of course, Jesus. You can see that our physical development as a species went along "normally" for thousands of years, and then suddenly we get this unexpected spiritual dimension appearing in *Homo sapiens.* It's interesting that Genesis suggests that the only way humans are distinguished from animals is in their spiritual qualities, the ways in which they are "like God." Consequently it is logical to suppose (and Genesis 2:7 suggests it), that God placed his image in an already existing animal, making him "a living soul."

Are you now suggesting that all this stuff you have been telling us is compatible with what it says in Genesis? Come off it!

Hey! You're making a huge assumption yourself! Why should the account in Genesis not agree? It's complementary to the scientific account, not a competing version.

But it is interesting to see how closely the account in Genesis does fit with the view of punctuated evolution that we mentioned before. Hebrew uses two separate words for making things: *bara*, which means to create from scratch, and *asah*, which means to shape something out of preexisting material, like a potter shaping a pot. Genesis usually uses the *asah* form, which is perfectly compatible with divine creation through natural processes like evolution. It only uses the word *bara* three times, and each time it is referring to a special creation: the creation of the heavens and the earth, the creation of living creatures, and the creation of humans. [15]

So where would Adam and Eve fit into this lot?

It would seem that, when the time was ripe, God selected an already existing member of the *Homo sapiens* species, and transformed him in some spiritual way, which Genesis can only describe as making him "in God's image." So while the Adam and Eve of evolutionary theory probably lived in Africa about 100,000 years ago, it's likely that the Adam described in Genesis, the person whom God invested with his spirit in a special way, was perhaps a Neolithic farmer some 10,000 to 12,000 years ago. This was when the "Sphere" decided that the time was ripe, that we "Flatlanders" had developed enough for him to be able to consider communicating with us directly, and gave us the ability to respond. We were now ready to receive a spiritual dimension, perhaps as an emergent property.

Adam represents the first person to have self-awareness, self-consciousness, and consciousness of God. He also symbolizes the first person to really have the freedom to choose a course of life. As we know, he chose his own way, and therefore lost close communication with God.

Now we don't know whether Adam and Eve were just one couple, or whether they represent an emerging family to which God gave the gift of a spirit. After all, the book of Genesis isn't a scientific treatise about anthropology, and does not have to be interpreted completely literally. What it does do is set out the basic fact that God created Adam with a spiritual dimension in the physical body of a Middle Eastern farmer.

How can you set a date and a place for a story that old?

Well, it is clear from the details in the story that Adam was supposed to cultivate the ground, which makes him a farmer, which puts him in the early Neolithic period, at the end of the late Stone Age as hunter-gatherers began to settle down. The start of an agricultural way of life coincided with an increase in population all over the world, but archaeologists have found the first traces of this way of life in Israel and the Middle East. They have even found evidence of it near the headwaters of the great rivers of Mesopotamia—and Genesis mentions the River Euphrates as one of the rivers of the Garden of Eden.

But I wouldn't get too hung up on the details of identifying an exact place or time. It is as an allegory or parable that the story of Eden and of Adam and Eve is most revealing.

How did this "parable" get to us?
After all, Adam and Eve didn't know how to write.

Right. We know almost nothing about the earliest days of humanity until the words and stories of the oral traditions of these Neolithic farmers began to be written down. One of the places where this happened was in the Middle Eastern countries of Israel and Judah. The tradition handed down from the early hunter-gatherers of the Paleolithic or Old Stone Age was preserved in the first chapter of the book of Genesis. A slightly different version is preserved in chapter two, which seems to reflect the experience of Neolithic farmers, who cultivated the land. Presumably these stories were retold to each generation until, with the arrival of a written script, the accounts of the beginnings of morality and of man's search for God were written down in a version that combined the separate accounts from Northern Israel and from Judah.

But why were they written down as if they were intended
as history or as science? Were they just fireside stories?
Or did they have some other purpose?

Genesis was written to teach essential truths about God that we could not find out for ourselves, and the early chapters are intentionally symbolic. It opens with the beautiful story of the Spirit of God overcoming chaos, light penetrating darkness, heaven differentiated from earth, and with humanity, represented by Adam and Eve, living together with God in love and harmony. Their developing awareness is personified in the form of the serpent—the dark depths of man's awakening consciousness. This awakening is the focal point of the Genesis story.

An evil being (we'll talk more about this later) takes the shape of the serpent and tempts the recently new man and woman, not to hate or reject God completely, but just to turn away from total dependence on him. "To go into business for themselves." They took it's advice, and this led to their experiencing guilt and shame for the first time, and to their needing to sacrifice if they were to worship God. In a sense, their newly awakened life "dies." *Homo sapiens* became self-conscious: they hid their sexuality from each other, and attempted to hide their self-awareness from God.

Genesis is no exposition of anthropology, and was never intended to be taken literally. It uses the images of a prescientific cosmology to express abiding religious insights. So does the Bible; it is God's self-disclosure in the life of one nation, Israel, and subsequently in one person, Jesus.

So what is the main message that God is disclosing about himself in the opening chapters of Genesis?

The first one seems to be that he played an active role in putting this world together, and from his continuing interest in the man and woman we get the impression that he is interested in conserving and sustaining it too. He also tells us that the world is essentially good, and that there is more to the universe than mere materialism. Those seem to be his main messages about himself.

If that is his main message about himself, what's his message about us humans?

There are a couple of messages about us too, and about the relationships between men and women. For instance, its says that men and women need each other. Oh, we can survive quite well without each other, as Adam did at first, but there is a loneliness, as Adam also found. God gave us to each other as companions. It's also interesting that God gave Adam only one wife. After all, at the time Genesis was written, your importance as a male was judged by the number of wives you had. But God's pattern was monogamy: one man, one wife.

And then there is the matter of rest. God knew it wouldn't be good for us to be workaholics, spending every day immersed in work. So in Genesis he reports on a seven-day cycle, six days of work, and a seventh on which we have time to rest and enjoy our relationship with each other and with God. Of course, Adam and Eve's disobedience rather spoilt that, and it also explains the "sickness of spirit" that seems to afflict humans. Even the good things we try to do often seem to have bad consequences, and too often we actually want to do things that we know are bad for us or for others—but we go ahead and do them anyway. Genesis gives us some clues as to why this is so.

I suppose that how we developed spiritually makes some sense and science doesn't actually deal with those things. But I am still puzzled about what you mean when you talk about human consciousness as being evidence that God has given man a spirit, or a soul, or whatever you want to call it. What do we actually know about our minds?

You want a one-paragraph answer to that question! I think it had better wait till our next session, when we can talk it over in far more detail.

8. Consciousness and What Comes After

Consciousness brings the mind alive; it is the ultimate puzzle to the neuroscientist. It is your most private place. —SUSAN GREENFIELD[1]

By now, the students were well aware that there were still gaps in scientific knowledge, and that the triumphalism of science was never valid. They had also begun to recognize that there was no fundamental conflict between science and the nonliteralism of the poetic, prescientific vision of creation. It was time to move on to examine deeper issues of the human mind and spirit, including the extraordinary mystery of consciousness.

What exactly do you mean by consciousness?

Consciousness is each individual's awareness that they exist, that others exist, and that there is a world beyond themselves. But that could also apply to animals, who know that they exist, that the pack exists, and that food exists out there. So the term has to refer to more than that: it has to involve the awareness that I, and the things around me, have a past, a present, and a future, and that I can know something about a past that I may or may not have experienced, and can anticipate and affect a future in which I may or may not be present.

What you are saying implies that only humans have consciousness, but don't animals also have it? Some animals seem to have memories, and the ability to make plans.

The trouble is that the word itself is probably too broad. We need to distinguish levels of consciousness, with the lowest level simply being any sign of brain activity. (Although Teilhard de Chardin famously suggested that consciousness begins even in stones![2]) Neuroscientist Susan Greenfield of Oxford University speaks of a spectrum of consciousness, a gradually increasing awareness as the brain itself develops—perhaps like a dimmer switch that can be slowly turned higher. Chimps have the switch set lower than humans, and a fetus has a lower level of consciousness than an adult.[3]

What would you say are the signs of a higher level of consciousness? The sort that you think humans have and animals don't?

I think that the ability to verbalize our thoughts is one of them. Human communication isn't just issuing instructions to one another or simply checking that the other person is there; it also involves communicating abstract thoughts to each other, the sort of thing that Descartes was struggling with when he said, "I think, therefore I am." A subset of this is the ability to use my imagination to create things that did not exist before, in the form of art, music, poems, and stories.

When we started talking about this, you referred to consciousness as a mystery. Why?

Well, to start with there are the simple mechanical or biological problems. How does the brain work? We still know very little about its amazing complexity. But even if we were to make rapid strides in understanding all the physics and chemistry involved, we would still be faced with the problem of understanding how physical processes in the brain give rise to our own subjective experience. Where does individual consciousness come from? Why are human beings so different from one another? Why is our consciousness so deeply rooted in our subconscious, which is like the submerged portion of an iceberg? Why do we have a level of consciousness that no other animal seems to reach? Why are we the only animals who create art and music? Why are we the only animals who worship? At our last session, we saw that Genesis gave answers to some of these questions, but science is still looking for more detailed explanations.

But surely neuroscientists have some of the answers to these questions?

Well, Professor Susan Greenfield calls the mind "the ultimate puzzle to the neuroscientist. It is your most private place."[4] She would say that they have a few of the answers, but by no means all. For example, neuroscience can explain how some psychological problems are caused by problems in the chemistry of the brain. And some neuroscientists think that they will eventually be able to prove that consciousness is nothing but a product of chemistry and physics.

But many others aren't so sure. Euan Squires, for example, a theoretical physicist at Durham University, identifies the central problem as the relation between physics and the conscious mind and stresses that all scientists acknowledge "the mystery of consciousness."[5] The trouble is, neuroscientists still can't give us any explanation of the physical basis of consciousness. It isn't located in any one part of the cortex of the brain. And they can't explain how the brain is related to quantum physics.

Huh? I never even thought of linking the brain to quantum physics. I suppose logically there should be a link—but can you explain what you were talking about?

Remember that way back in one of the early sessions we talked about observer-centered reality? It was only when something was observed that the wave function collapsed. Only at the moment of intervention by a conscious observer is it settled what the answer is going to be. So we have the peculiar dilemma that we have in "mind" (or rather in "consciousness") a nonmaterial "thing" that is, "on the one hand, evoked by the material world, and, on the other, can influence it."[6]

Could this have anything to do with the different dimensions we were talking about when we were discussing physics?

It might well, but few neuroscientists are exploring this possibility. They are still trying to unravel the biological mechanisms of the body, and don't yet look beyond that. A few, such as Sir John Eccles, see the mind or soul as distinct from the body.[7] J. R. Smythies has proposed that the universe may be made up of a sort of honeycomb pattern of a common physical world surrounded by a number of private mental worlds.[8] Professor C. D. Broad suggests something similar when he argues that physical space and the space of the mind are different sections of a single multidimensional space. The mind is extended, but not through physical space.[9]

It is useful to talk about the soul as the mysterious meeting point of a body with a spirit (i.e., soul puts man one up on a ghost in a machine). Death then means separation of body and spirit, and resurrection means the soul is resurrected—a new body plus the same spirit. Only by having a spirit that can survive without a three-dimensional body can one validly maintain identity between the dying soul and the resurrected one.

So it seems as if consciousness is both the most familiar thing in the world and the most mysterious—even more mysterious than quantum physics!

Yet every moment of every day, we take this amazing thing for granted. I find it very difficult to explain without the existence of a Creator, who is both within and outside our physical dimensions.

If consciousness is this amazing, what happens to it when we die? Does it just evaporate?

That's a question we humans have been trying to answer for centuries, with some people saying there is definitely some sort of afterlife, while others say that at death we just cease to exist. There is some interesting evidence from what scientists call "near-death experiences" that supports the idea that there is something after life—Melissa here had an experience like that.

But of course, the only way to know is to find someone who has really been dead—dead long enough for the brain cells to completely stop functioning, and then question them about it. That's difficult to do! Mediums and channelers try to do it, but there is quite a lot of fraud associated with some of them, and with others, we don't really have any proof of whom they are talking to.

The only source we have for information about this is the one guy who did come back after having been dead for a few days, and hung around in this world long enough to convince a lot of people that he was the same person who had died physically, and that he was thoroughly alive again—alive enough to eat with them, walk with them, even be present with them at a lakeside barbecue. Of course, he wasn't *exactly* the same as he had been before. He could do things he hadn't been seen to do before, which suggested that life after death was more interesting than life before it!

You're talking about Jesus, of course. And I suppose that does prove there is some kind of afterlife. But the Christian view of the afterlife seems so boring, hanging around on clouds strumming harps. What's the point of it?

Where do you get that picture of a Christian afterlife? From cartoons and Victorian sentimental pictures? You certainly won't find anything like that in the Bible, where the afterlife is both a lot better and a lot worse than that image suggests. John Polkinghorne, a particle physicist turned Anglican priest, describes the Christian hope of resurrection, as the reconstruction of a "pattern, dissolved by death, in a new environment of God's choosing."[10] We don't have many details about it, and the images in the Bible are quite clear about the fact that they are just images, symbols used to express the inexpressible.

But that new environment of God's choosing could be either heaven or hell, couldn't it? Where are those places? Are they really places?

Do you remember living in Flatland? When the Sphere tried to explain three dimensions to people who lived in only two dimensions? Where would they have thought that heaven was located? It would have been very difficult to explain to them, wouldn't it? We face the same problem when we talk about heaven and hell.

But there is one interesting comment from the Sphere who came to our Flatland. Jesus said that "the kingdom of heaven is within you."[11] What does this mean? Well, it can be interpreted as meaning "within your grasp" or "waiting for you to sign on,"[12] but at the very least it means that heaven isn't just something "out there," but also involves our potential to know the Creator for ourselves, through Jesus, and to experience something of his world.

But in Flatland we said that the Sphere was good and kind, and wouldn't just destroy Flatland. Yet you say that the Christian Creator can send people to hell! How can he do that? What do Christian's mean when they speak about hell? Who gets sent there? Sinners? But who are sinners?

God's definition of sinners isn't quite the same as the newspapers, or yours. Jesus showed kindness and forgiveness to the scandalous adulterer, the outcasts, the poor, even hated tax collectors. On the other hand, he told a couple of stories that suggest that the people who are sent to hell are the self-righteous pillars of respectability in their ivory towers of pride, who are therefore unable to love.[13] They think they have earned their place in heaven by their legalistic concern with the letter of the law.

Sin is essentially *not* a transgression against a rule, but rebellion against a person—Jesus. The essential malice of sin consists of a refusal to return God's love. The greatest of sins is to be conscious of no sin! (Or not even to have a sense of need—which would open the door to forgiveness).

So if we're wrong about what constitutes sin, are we also wrong about what hell is like? Isn't it a sort of torturer's dream world, the sort of places sadists would like to work in? And it lasts forever and ever.

When we were talking about heaven, I quoted what Jesus said about the kingdom of heaven being within you. So the reverse is probably true too. Hell is not just an external experience of suffering, but as John Habgood reminds us, internal torment, produced by our "unwillingness to open ourselves to love."[14] Michael Ramsey states that hell is if our free will were to go on preferring selfish isolation to being joined with the love of God, and warns us that "to fail to respond is to slip into denial, which means the rejecting of our privileges."[15]

The image that the Bible most often uses for hell is a fire that consumes what is unworthy of entering God's presence. The book of Revelation says that at the very end, hell itself (Hades) will be thrown into the lake of fire,[16] signaling an end to all suffering in the new heaven and the new earth where there will be no more death. So it seems that those who reject God will not be in torment forever.[17] The picture of the everlasting torture of the damned is a symbolic way of speaking about how terrible it is to reject God and therefore be rejected by him. David Jenkins, though agnostic about hell, says that if you persist in moving in the opposite direction to God's love, "you may do yourself out of a life and come to nothing at death."[18]

But in rejecting people, God does judge people—that can't be denied. Archbishop George Carey[19] and many other Christians are convinced that part of that judgment is exclusion from God's presence.

What did Jesus himself say about hell?

Despite telling us how important love is, which is the part we all like, he also said that those who do not or will not forgive will be handed over to the torturers.[20] At least part of this torture is the internal torment and bitterness that goes with being unwilling, even unable, to forgive, compared with the peace of soul and spirit in genuine forgiving and receiving forgiveness.

But how can love and forgiveness be that important? Generally I get on OK with people, but there are a few people I am never going to be prepared to forgive because they really hurt me.

The trouble is, that those are the only people we need to forgive—the others haven't done anything much that needs forgiveness! Think of love and forgiveness as being a bit like a muscle. If you choose to lie in bed all day, every day, your leg muscles will gradually atrophy. They'll wither away and become useless, and won't be any use when you do need them. The same happens with love and forgiveness: if you don't use your capacity to love, it withers away. Eventually you'll end up in a position where you can't love God, or other people, and the only person you do love is yourself. And then what is there left for the Sphere to connect to? You'll only want your world, and not have the slightest interest in the universe that God wants you to experience. You'll cut yourself off from God and from heaven.

You are basing an awful lot of what you are saying here on what Jesus said. And you said you were prepared to believe him because he came back from the dead. But how do we know that really happened? It seems unlikely to me.

First, let's deal with the problem that any resurrection looks unlikely. Think about Flatland again. When we talked about the Flatlanders killing the Circle, we agreed that they couldn't really damage the Sphere. Because the Circle was a part of the whole Sphere, it would have survived, no matter what they did to it. We also agreed that the Circle would probably have come back to reassure the Flatlanders who had believed in it that it was still alive and that what it had said was true.

If this seems logical in Flatland, it should be equally logical here. And it is equally logical that the Flatlander's response to the returned Circle would at first be one of disbelief. Because they didn't fully understand his relationship to the Sphere, they were sure he was dead, and dead Flatlanders weren't in the habit of coming back to life either.

Jesus' disciples reacted just like the Flatlanders. Those who met him reacted with unbelievable joy, but the rest of his followers didn't believe the first reports they received, and thought the women who claimed to have seen Jesus were suffering from delusions or had seen a ghost. Thomas even insisted that all the other disciples must be mistaken when they told him they had seen Jesus.

But the reports of sightings kept coming in, and the behavior reported wasn't very ghostlike. Two of his followers rushed in claiming that while they had been walking home to Emmaus, very depressed about the failure of all their hopes and dreams, Jesus had joined them and walked all the way home with them. They hadn't recognized him at first—after all, it was getting late, they were wrapped up in their misery, and he was the last person they were expecting to see. But they were grateful for his companionship, and for the things he said about Old Testament prophesies that might be coming true. And when they invited this "stranger" in for dinner, he suddenly did something in such a characteristic way that it was as if scales fell from their eyes and they recognized him. He had left them shortly after that, but they were so excited that they rushed the whole seven miles back to Jerusalem to tell the others.[21] You don't walk fourteen miles in one day unless you are pretty convinced that the guy you met was real, not a hallucination or a ghost.

But I still think it's odd that they didn't recognize him—after all, they knew him pretty well, and should have recognized his voice at least.

Yes, and we find this uncertainty in quite a few of Jesus' appearances. The disciples knew that it was Jesus, but it wasn't a case of his physical body having been resuscitated. He did things he hadn't done before, like joining his eleven disciples in a room where all the doors were locked.[22] It is as if, in his resurrection, the physical and spiritual became one: the human person was reconstituted, whole and entire. And the disciples' response to him was a mixture of joy and awe.

But, of course, this complete blending of the physical and the spiritual means that at times the spiritual dimension is stronger than we are used to. After a few months, Jesus stopped appearing in a tangible way, and when he appeared to Paul on the road to Damascus some two years later, it seemed more of a spiritual than a physical encounter. Maybe that was why Paul later makes the distinction between our physical bodies and our spiritual bodies[23] and "ransacked his vocabulary"[24] to find appropriate language to express this utterly new dimension of his encounter with Jesus.

C. S. Lewis tries to capture something of this difference between the physical and spiritual in his book *The Great Divorce*,[25] which describes a day trip to heaven. The narrator's wonder at first stepping out into the beauty of heaven is tempered by sharp pricking sensations in his feet. When he looks down, he discovers that the grass of heaven is actually growing through his feet: it is more substantial than he is. He appears to be a ghost by comparison with the reality of heaven. In the same way, the resurrected Jesus somehow seemed more "real" than the world around him. It was this type of difference that led Thomas, the most skeptical of the disciples, to finally acknowledge Jesus as "My Lord and my God."[26]

The accumulated weight of the evidence is compelling. Jesus was alive from the dead and making himself known to them "in a mode of existence that transcended the limitations of time and space."[27] The accounts of the resurrection certainly show us Jesus freed from the boundaries of merely local time and space; he clearly transcends the four dimensional spacetime in which we think we live. He is living in a new mode of life, with a new body that is really a new creation.

If I accept that Jesus did come back from the dead, does it mean anything more than that there is life after death?

Well, removing the fear of death (although also making us aware of judgment after death!) is no mean feat.

But the resurrection doesn't just affect how we die, but also how we live. Look at the disciples. They had been devastated with shame and disappointment, terrified that they might share their Master's fate. But once they were convinced that Jesus was alive, there was a dramatic change. Suddenly they had joy, a conviction that they had a task to do, an understanding of what Jesus had been trying to teach them all along, and a confidence that through the Holy Spirit Jesus was with them and would never leave them again. The prospect of persecution didn't bother them any more. And that wasn't just a phenomenon for the disciples then—we see it continuing in the worldwide spread of the Christian faith, and in the joyous faith of believers in China, South America, and Africa in the face of widespread persecution.

But how can I force myself to believe this?

It's not really a case of forcing. We have looked at some of the evidence, and you can look at more of it if you want to.[28] The evidence for the resurrection is strong.

But earlier I made the point that the resurrection itself is both physical and spiritual, and the same goes for our response to it. Once you have looked at what we could call the physical evidence, you also have to listen to the spiritual evidence, listening for the resonance of this story within yourself. It is the Spirit we allow within us who recognizes Jesus. As someone has said, "The Lordship of Jesus is not deduced but encountered."[29]

And finally, you have to recognize that few things in this world are certain in advance. You may well find that any lingering doubts you have are resolved by experiential reality, in a new dimension of seeing. James

Dunn of Durham University has pointed out that our encounter with the risen Christ often becomes clearer after a revelation of assurance of forgiveness and of being commissioned for service by Jesus.[30]

Quite often the question is not so much do you believe, but do you *want* to believe?

9. Miracles and Missions

In the Bible God is always present, however strangely, and "deeds of power" are seen as special acts of a present God rather than as intrusive acts of an absent one. —TOM WRIGHT[1]

OUR LAST SESSION had ended with my challenging the students to ask themselves whether the questions they were asking were real questions, or merely attempts to fend off Jesus' claims. When we met again, their questions were sober and focused.

You keep referring to the Bible, but I have real difficulty in taking it seriously. All those miracles . . . it sounds like a string of fairy tales.

Actually, if you read the Bible, you might be surprised at how few nature miracles you come across. There are a few times in the Bible when there seem to be outbreaks of miracles but, in general, it's pretty realistic and doesn't read anything like a fairy story.

But even if there aren't that many miracle stories, the ones that are there are enough to make me suspicious of the whole thing. You've made a fairly convincing case for there being some sort of Creator, but why should that Creator go around tampering with the physical laws of nature?

Actually, very often he doesn't tamper with them. He respects them, and uses them. We've already seen that when we talked about the anthropic principle and the way he shaped this world to be suitable for life.

But let's actually look at one of the miracles described in the Bible, the one where the Israelites cross the Red Sea. You've probably seen the Hollywood version, with walls of water rearing up on either side of a miraculously dry sea floor while the Israelites cross, and then coming crashing down on the Egyptians. It looks impressive—and very contrary to the laws of nature. But look at what the Bible actually says happens: the place where they crossed was the Sea of Reeds, at the northeastern end of the Red Sea. Sure, Moses told the Israelites that God would deliver them and held his staff out over the water, but the water didn't roll back immediately. Instead, a strong east wind blew all night, blowing the shallow waters back and turning the marshy sea into dry land that the Israelites could safely cross. By dawn the wind had dropped, and the water started to flow back. The chariots of the pursuing Egyptians got bogged down in the mud, and as the water returned to its normal level many of the Egyptians drowned.[2]

The miracle is in the timing, not in any rearrangements of natural processes. God demonstrated his care for his people and his response to specific prayer without having to violate scientific principles.

I suppose that's plausible, and I have read similar sorts of explanations for the plagues in Egypt too. But what about Jesus? I get the impression that he walked around scattering miracles right and left. He seems to have ignored the laws of science.

If you actually read the Gospels, you'll see that Jesus didn't spend all his time doing miracles, and you'll see that he was very concerned about the right and wrong use of miraculous power.

For example, once after Jesus had been fasting for a number of days, a voice inside him started nagging, "You're hungry, aren't you? *If* you are the son of God, with all these miraculous powers, why not just turn these stones around here into bread. Nothing wrong with having a bit of food, and it would give you a bit more confidence that you are who you say you are. Come on, why not do it?"

But Jesus recognized that that voice was pushing him in the wrong direction, and insisted that it wasn't his job to feed himself, but to listen to God. And God had told him that he was his son, so there was no need to go running around trying to prove it.

And then the voice began again. "OK. You're right to be secure in who you are. But how will everybody else know? They'll just have your word for it. Why not try and do a really spectacular miracle, like floating down from the top of the temple in front of a whole crowd? That'll convince them you're somebody special, and it won't involve you or anybody else getting hurt. There's a lot to be said for that, you know."

But once again, Jesus recognized that miracles were not meant to be stunts. So when he does eventually get around to performing miracles, he uses them only to benefit others, not himself, and he often uses them almost like parables, to drive home a lesson.

So you are admitting that Jesus did do miracles that can't be explained away by science?

I don't think I "explained away" the other miracle either. The timing was still miraculous. And I also think that somebody who can come back from the dead doesn't really need to obey the same laws of nature that we do—even more so if you think that he was the one who designed the rules in the first place.

But there's something else you should consider too. You're creating a false distinction between the natural and the supernatural. Remember that we talked about the fact that the chair you're sitting on looks solid, but is actually composed mainly of empty space and force fields? In that case, doing something miraculous is really just a case of rearranging that energy into a new pattern. When he healed sick people, Jesus was somehow able to rearrange a damaged pattern into a healthy one. We don't know how he did it, but there is pretty good evidence that he did, including the fact that miracles still happen today (though not as often as we would like!).[3]

I suppose that the idea that he was rearranging energy patterns might explain how he could turn water into wine and healing people, but what about other miracles that seem to involve more than that? Isn't he supposed to have fed 5,000 people with just five loaves and two fish?

You know, you might find miracles a bit less bewildering if you thought of them as "special acts of a present God, not intrusive acts of an absent one."[4] If we spend hours arguing about the details of *how* they were done, we are ignoring the much more important question of *why* God chose to do something special at that particular time. Remember that I said that Jesus often used his miracles like parables, symbols to underline a particular point. That's what all of those miracles are really about.

I don't get it. Can you explain that a bit more? What did those miracles symbolize?

OK, let's start with the feeding of the 5,000. People can argue about whether Jesus created new bread or whether he was able to encourage everybody in the crowd to share their emergency food supplies, but whichever way it happened, it sent an important message. If he did it the first way, he demonstrated his ability to give people what they need. It underlined his words "I am the bread of life . . . He who comes to me will never go hungry."[5]

If he chose to do it the second way, he was still making that first point, but was also driving home the lesson that there was no need for some people to be hungry if we all cared for one another and shared what we had (a lesson we could do with hearing again today, when so many people live in poverty or as refugees). Jesus dramatically illustrates the point that one of my friends makes: "The Lord adds to whatever you have if you give it to others."

And the water to wine story? What's the message there— "LET'S PARTY!"

You hit the nail on the head! That was part of his meaning. After all, the coming of the Messiah, who Israel had expected for centuries, was something worth celebrating. But there was more to it than that. The jars that held the water were ritual containers, and he was making the point that he could replace boring water, the legalistic ritualized approach of Judaism, with the best wine, the excitement of participation in the kingdom of God. No wonder the host was told, "You have kept the best wine until last."[6]

You're saying that Jesus replaced Judaism. But there are still plenty of Jewish synagogues around today. So he didn't manage to replace it, did he?

I didn't say that Jesus had replaced Judaism; I said that he had replaced the legalistic ritualized approach to Judaism. But although many Jews followed Jesus, not all did, and that is why the Jewish tradition continued to develop. So you'll find that modern Jews and Christians share the same moral code, the Ten Commandments. They share the same holy book, the Old Testament, and the same admiration for the spiritual wisdom of prophets like Amos, Hosea, Isaiah, Jeremiah, and Ezekiel. Where they differ is that today's Jews also show enormous reverence for the oral tradition, known as the Mishna, and the rabbinic commentaries, which together with the Mishna make up the Talmud.

If you talk to many modern Jews, you'll find that they think of God as an awesome but remote being, rather than a heavenly Father who loves us deeply (which was what Jesus taught us). Orthodox Jews worship him by strictly obeying a set of intricate laws, although Reformed Jews have discarded some of these as irrelevant to everyday life.

And, of course, the most important difference is that today Jews do not regard Jesus as the Messiah, the one who the prophets had said was going to come. But it's worth remembering that all of Jesus' first disciples were Jews. And today we are seeing increasing numbers of Messianic Jews, that is, Jews who acknowledge Jesus as Lord and worship him through the Jewish faith.

I like that bit about worshipping him "through the Jewish faith." Isn't that true for all faiths? Aren't all religions just different paths to the same God, and just as good as Christianity? After all, you said that William James found some common elements in all religions.[7]

Yes, there are things that all religions have in common, but there are also some big differences between them. If you want to know about other faiths, including the faith of Marxists and atheists, I recommend talking to people who are immersed in those faiths, who live out their own religion in sincerity.[8]

I've found it really worthwhile to listen to men and women of different faiths, until I can begin to see the world through their eyes.

I've a friend who is a Sikh and always wears a turban, but I've never asked him about what he actually believes. Have you spoken with any Sikhs?

Oh, yes, I've a Sikh friend who teaches physics. From talking to him, I've learnt that he believes that, without the inspiration that comes from devotion to the one God, people live in illusion and are dominated by five evil impulses: lust, anger, greed, attachment to worldly things, and pride. But he doesn't have much interest in the combination of Hindu and Muslim theology that underlies the Sikh faith. What is much more important to him is making sure that he retains the physical signs of his religious purity, or *khalsa,* which include not cutting his hair or beard, and always wearing a comb, a steel bracelet, short trousers, and a double-edged dagger.

My Sikh friend comes from India. I was surprised when I met him, because I thought Indians believed in the Hindu religion. How do Hindus differ from Sikhs?

Well, for one thing, the Sikhs believe in only one God, whereas Hinduism has many gods. However, some Hindus, like another friend Mr. Malhotra, do believe in one all-powerful mysterious God going by the names of Brahma-creator, Vishnu-preserver, and Shiva-destroyer. But there are also levels of other nationally recognized gods such as Krishna, and a lower level of village gods, a vast army of spirits.

Castes are another important aspect of Hindu life, with everyone born into a caste that reflects how well they behaved in a previous life. If they live well in this life, they may hope to be reincarnated in a higher caste. Yet many modern Hindus are becoming increasingly uncomfortable with this system, and the discrimination it so easily leads to.

But I thought that everyone was heading off to India to learn about Buddhism. What do you know about that? What do Buddhists say about God?

Buddhists don't actually have much to say about any supreme God. In many ways, their faith is almost atheistic. The main focus is on the process of trying to become perfect, earning salvation by following the eightfold path that emphasizes self-denial, right conduct, and the like.

Like the Hindus, Buddhists insist that all significant actions have their consequences, often worked out in reincarnation. Salvation is described as the ultimate escape from the cycle of death and rebirth, which is achieved by subjugating all desires and achieving the state where one has no desires at all. The Buddhist looks for peace in what in some ways could be described as a sort of suspended animation.

What about Islam? How does it see God?
To judge by the news these days, there are a lot of people who are very committed to it, but some of them don't seem to care much about making others suffer.

The prophet Muhammad portrayed a remote and distant God, rather than a God of such infinite love that he wants us to call him "Father." It seems that some aspects of Islam encourage men to go fearlessly into battle against infidels (literally, the faithless) and assure them that if they die in battle they will go straight to paradise. Suicide bombers seem to be inspired by this belief in a jihad, or "holy war." This is more aggressive than the Christian idea of martyrdom being a refusal to compromise one's faith under persecution.

Yet there are good things in the Islamic faith. God is seen as the source of life and they have great respect for his natural authority, and for the written authority of the Qur'an. In fact the very word *islam* means "surrender," accepting what Allah says and does. One of the ways they demonstrate this surrender is obeying the requirement that they pray five times a day and fast for the whole month of Ramadan.

Muslims would say that Allah revealed his will through prophets like Jesus, but that the final revelation was through the prophet Muhammad. They deny that Jesus actually died on the cross, and so they would say that there was no need for him to come back from the dead.

Listening to what you say, it seems as if there is quite a difference between what different faiths imply. They don't sound as if they are all saying the same thing at all. They may agree on a few points, but there are huge things that they disagree on. So it seems as if we really do have to choose between them.

Yes. It is good to recognize that there are similarities between our spiritual experience and that of others, but it's also painful to recognize real differences and disagreements, and to realize that we can't just coast along, avoiding any choices.

So what do you think is the biggest difference between them? Or maybe the biggest similarity?

That's a tough question—but you may have accidentally hit the nail on the head in the way you asked it. The similarities and the differences are linked!

William James pointed out that all religions emphasize the need to make a proper connection with higher powers.[9] In almost all religions, including Christianity, the way to make this connection is often presented as being submissive to a traditional code. Islam incorporates that in its name, Judaism has strict rules for kosher eating, the Sikhs have to carry distinctive symbols, the linked Buddhist must obey the eightfold path. But this type of rule keeping can be deadening. As someone said, speaking about the Christian church, a moral code based on the Ten Commandments may well lead to a "hardening of the oughteries"!

When you have this type of system, you often end up with people having more faith in a system of religious doctrine or in a set of rules than in God—and that applies across all religions.[10] Not that the rules are always bad. Bishop Ambrose Griffiths, a Catholic bishop, has pointed out that laws like the Ten Commandments are "not just laws laid on us, but descriptions of reality. This is the world as it is and we ignore them at our peril."[11]

Keeping the Ten Commandments may look quite similar to the rules you will find in other faiths, and may look moderately easy to keep (with some exceptions, where we can normally rationalize our failure to do so). But then we get the summary of them: Love God above everything else, and love one another as he loves us. Then we realize that we haven't a snowball's hope of coming close to keeping them!

And that I think is where Christianity parts company with other religions that are partially true but inadequate. They tell you how you ought to live, and then you're on your own. Jesus came to tell us that God cares about the fact that we struggle to obey him. Not only that, he offers us the gift of his Spirit to work in us and transform our lives, giving us the power to serve him better. No other religion includes the presence of the Holy Spirit.

Moreover there is no force involved. Jesus never suggests that there is some sort of rigid cultural rite that we *must* perform to demonstrate that we are his followers. No fixed prayer times, no ritual daggers, no abandoning the world—and no social pressure to force us to follow him. All he does is ask us to recognize that we cannot serve God properly on our own, and ask us to freely accept the work of the Spirit of the resurrected Jesus in our lives.

Finally (and this may be a new insight of this book) the Christian hope of heaven and a new earth, understood as a bodily resurrection in more dimensions, dovetails remarkably well with the new scientific worldview.

But the Ten Commandments were only given to the Jews, and Jesus only went to Israel. What were people in other areas to do?

God's spirit has ensured that every culture has some awareness that there must have been a creator—even our scientific culture, as we saw when we talked about the anthropic principle. Religions developed in response to this awareness, and that is why Bishop John V. Taylor can write: "In Jesus each religion will be brought to fulfilment in terms true to itself, through crisis and conversion."[12] He foresees a new Jesus-centered Hinduism (Mahesh Chavda is an example of the direction this might take)[13] and a Messiah-centered Islam. He can have this vision, because Jesus is both universal and unique.

What do you mean "both universal and unique"?

Well, Jesus is unique in the sense that he lived in a particular place at a particular time in history. He also has a unique relationship to God that no one else can possibly duplicate (for an analogy, think of the relationship between the Circle and the Sphere in Flatland).

But he is also universal, in the sense that what he had done and what he offers applies to everyone, and can be understood by everyone—sometimes in the most surprising places. For example, the remote Sawi culture of Irian Jaya (New Guinea) is about as far removed geographically and culturally from the area where Jesus lived as it is possible to get. The Sawi valued cunning, and had great admiration for anyone who could really deceive an enemy, luring him into a sense of safety before killing him—and then using his trophy skull as a pillow. Judas was a hero to them! Not surprisingly, they were also engaged in bloody long-term conflicts with neighboring tribes. However, a missionary who went to live among them found that there were some things that he and the Sawi had in common. They shared a belief in the supernatural world and in the importance of interaction between the supernatural world and this one (although they disagreed about how this was done). But it was difficult for him to get much beyond this agreement, until he discovered how the Sawi concluded a peace treaty. They did this by exchanging a peace child, sending a young child from their tribe to live with the other tribe as a guarantee of their good faith. Don Richardson was able to tell them Jesus was God's Peace Child, effecting reconciliation between them and the supernatural creator. Suddenly, Christianity was not an alien concept, but rather the fulfillment of the best in their own culture.[14]

That's a bit like what we did with the Flatlander analogy, isn't it? Although that one was real, whereas ours was just imaginary. Are there other stories like that out there?

Yes, Don Richardson, the missionary who discovered this link was so excited by it that he went looking for other examples of what he called "redemptive analogies" in other cultures. And he found plenty of them. For example, the concept of one supreme God has existed for centuries in hundreds of cultures throughout the world, no matter how many minor gods there appear to be in the religion. Other groups have had legends about white-skinned strangers who will bring them a book their forefathers lost so many centuries ago—a book that tells the secrets of the supreme God.

Don Richardson found similar clear redemptive analogies in groups from Ethiopia, the Hawaiian Islands, North and South America, Central African, Thailand, India, and Korea. He was forced to conclude that the Christian message was unique in having had "an inside track" already laid for it in the belief systems of thousands of very different human societies.[15] But he also notes that missionary breakthroughs often came through the missionaries' respect for folk religions, rather than in spite of them! The really dangerous situation was where formal Christian religions were superimposed on folk beliefs before the Gospel had been understood.

But if God could give them all little bits of his message, why couldn't he give them the whole thing? Why did he just give that to the Jews?

That's another good question, and I can only guess that part of the answer relates to where the Jews lived, right at the crossroads of three continents. If God was to come to earth as Jesus, he could only be at one place at a time, but at least it was in a place from which the news about what had happened could easily spread to other areas.

As to why they didn't get more of the message in advance, I don't know the answer. But part of it may be that religion is not just a way of looking for God; it can also be a way of escaping from God, of hiding from God[16]—and this is as true of Christianity as of any other religious system. We like to hide behind rules that are easy to understand and that make life appear simple, rather than risking an encounter with a "Sphere" that operates in dimensions we know nothing about. Jesus challenged this rule-bound approach, and that was why the "religious" people wanted to get rid of him. Unfortunately for them, the resurrection proved that he was in the right!

So do other religions besides Christianity lead to God?

No religion leads to God, not even Christianity, without Jesus! As I said, they may even be an escape from the living God. "Religion" can take us into unreality, rather than getting us to tackle the problems of the real world.

"God is not the God of the religious or the religions, but One who is seen in the vulnerability and openness of Jesus."[17] He came to tell us that God actually likes us—in fact he loves us—and has gone out of his way to make a real relationship possible. His death made reconciliation and acceptance by God possible, and what he did was validated when he came back from the dead.

But isn't saying that Jesus is the one who has the truth really arrogant on the part of Christians?

Well, yes, except that we keep running into the fact of his resurrection, which no other religious leader has ever done, and into the ongoing work of his Spirit in the world. But at the same time, it's important not to feel proprietary pride that we are right and they are wrong. Jesus is not the property of Christians. Christians are no better than those who believe in other religions or claim to believe in none: "We all need a Saviour and there is no room for arrogance; we need to respect everyone, in love, as those who are made in the image of God."[18]

But how did you come to believe this for yourself?

My own experiments with other religions (including attending a Christian church) led me nowhere until I met Jesus and the Holy Spirit. Jesus gave me the assurance of forgiveness, the soul peace he offers. I found that, when I really came up against evil powers, only the power of the Holy Spirit could set me free from the darkness and depression (but I'll tell you more about that in later session).

I find this amazing, particularly when I consider that Abdus Salam, who has won the Nobel Prize for his work on the unification of forces, and who is a man of great humility and integrity, speaks of the eternal wonder of the spiritual dimension of life, uses the Qur'an to guide his life, but still awaits direct religious experience. And yet God through his Holy Spirit has given me very direct experience of himself. I find it difficult to believe, yet it's true!

10. Chaos and the Hidden Order

The whole history of science has been the realization that events do not happen in an arbitrary manner, but they reflect a certain underlying order. —STEPHEN HAWKING[1]

IN OUR LAST SESSION, we had looked at patterns of agreement and disagreement between religions, and at the paradox of God abiding by and flaunting the rules of the universe. Some of the students were still wrestling with how to make sense of the flood of information, scientific and religious, sweeping over them. It seemed an appropriate time to move on to a discussion of chaos theory.

Chaos means utter confusion, unpredictability, so how can there be a "theory" of chaos?

The fundamental point of chaos theory is that apparently patternless behavior becomes simple if you look at it in the right way. The classic shorthand way of talking about this is to speak of the "butterfly effect": the flapping of a butterfly's wing in Brazil sets off a tornado in Texas. What this means is that one small event somewhere can start a chain of events that has totally unforeseen consequences somewhere else in the world.

But in a way that seems kind of obvious. There must always be some sort of chain of events leading up to anything that happens. Why all the excitement about chaos theory?

One reason is that it gives us such beautiful pictures! Multicolored posters and Web sites with complex designs called fractals have caught the imagination of people who know nothing about the mathematics involved in producing them. The fractals have shapes within shapes that, however much you magnify the detail, still repeat the same pattern, as in the Mandelbrot set, which is just one example of a fractal image.

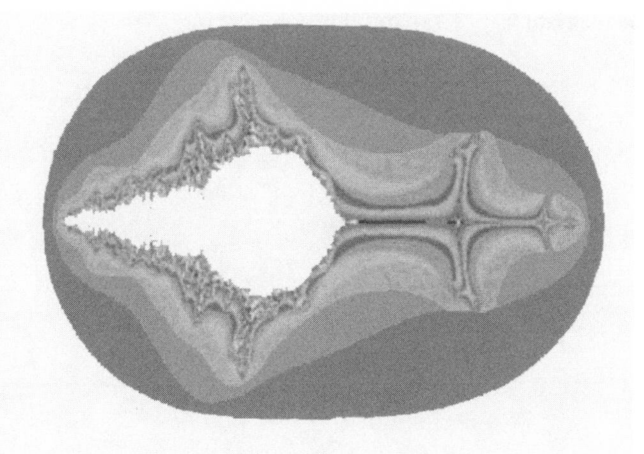

FIGURE 10.1 MANDELBROT SET

They are amazingly detailed pictures. It's as if you could dive down into the images and keep swimming down deeper and deeper into a sea of color where everything changes yet stays the same. How do they do it?

With mathematics, and powerful modern computers. A fractal is actually just a graphic display produced by a very simple but rigorous mathematical model. You start off with a simple equation, and then it grows and grows and grows. A fractal is a graphic example of the beauty and elegance of the mathematics.

I can understand why people like the pictures, but why do scientists get excited about them?

Partly for the same reason that you do, with the added value that they appreciate the elegance of the mathematics that produces the image. Scientists aren't aesthetic barbarians, you know! But even more, they appreciate fractals because of what they can learn from them. Think about it: when you first glance at a fractal, it appears to be an intricate web of patterns without much relationship to each other. But when you examine it closely, you start to see the repetition of forms until suddenly it falls into place and you recognize what the basic shape is. The apparently patternless behaviour becomes simple to interpret once you have the right way of looking at it.

FIGURE 10.2 FRACTAL IN NATURE

What this means is that what looks chaotic is actually highly ordered. And what scientists are constantly trying to do is to discern the order that lies under the apparent randomness of what we observe around us. They are hoping that simple mathematics, of the same type used to generate fractal images, will also help explain very complex phenomena, like the geometry of mountains, clouds, and galaxies. They hope to be able to use this to solve real-life problems. For example, if they could understand the mathematics of weather systems, they would be able to make accurate long-range weather predictions, which would be a huge benefit to people like farmers, who would love to get advice on whether they should plant a drought-resistant or a rot-resistant crop this year.

How do scientists set about doing this?

In the past, we did science by focusing on a little bit at a time, and trying to make sure that we understood everything that there is to know about that small part. That gave us a lot of useful information about details, but made it difficult to get the big picture. But the trouble was that the big picture was *so* big that there was no way we could handle it—until computers came along. But that didn't solve all the problems. Computers could crunch numbers for us pretty quickly, but they couldn't easily help us interpret the numbers they spat out. And then came fractal geometry and computer graphics, which could give us a visual representation, a map, of the overall shape of chaotic systems and help us to identify the hidden repetitive patterns that characterize the true nature of chaos.

In fact, the more scientists looked at these images, the more they came to realize that order and chaos coexist, entwined together in the same natural phenomenon. They found fractals in the shape of the edge of a leaf, the florets of cauliflower and broccoli, the outline of a tree, and in the course of a river. All of these showed the same degree of irregularity and the same complete structure whatever scale you magnified them to or reduced them to. The rules governing growth ensure that small-scale features become translated into large-scale ones.

More and more scientists realized that what we perceive as chaos is merely a reflection of a profound complexity that underlies seemingly random events.

You showed us a picture of a Mandelbrot set. Who was Mandelbrot? Why was the set named after him, or her?

Benoit Mandelbrot was a mathematical physicist who was born in Poland, studied in France, and eventually ended up working for IBM for many years, before becoming professor of mathematics at Yale. In 1977 he published *The Fractal Geometry of Nature* in which he tried to reveal the forms within nature—not just the visible forms, but hidden shapes "embedded in the fabric of the universe."[2] This book played a key role in the emergence of chaos theory because it opened scientists' eyes to the fact that fractals could be used as visual descriptive tools in the real world—used to describe turbulent motion, living organisms, blood vessels, and even the universe at large.

So where did he get his ideas from? I mean, how did this start?

About a century ago a French physicist called Henri Poincaré published a paper on celestial mechanics, that is, on the way objects in space move under the influence of gravity. He showed that gravity could produce extremely complex orbits and provided the first evidence of what is now called chaos.

In the 1960s, as computers started to become available, his geometric approach was redeveloped with techniques that included phase space, using more than the usual three dimensions. This multidimensional space could be represented in a fractal map.

How? If you've ever looked out of the window of a plane that is flying at several thousand feet, you'll have seen that the countryside below looks almost like a two-dimensional map. The mapping of multidimensional space works much the same way, except that it can give a three-dimensional image.

Why do fractals take the shapes they do?

Within chaotic systems there are definite deterministic laws that create three main patterns, each called an attractor. If the attractor is just a point, the pattern will return to one final resting place, like the bob on a pendulum gradually returning to rest. If the attractor is an ellipse, the pattern settles down into a periodic cycle and its future motions are easy to predict. This is the sort of pattern we see in the way the planets orbit the sun. If the attractor is strange, the things around it will seem to move wildly and erratically, yet will remain in some bounded region of space. Its behavior involves an equation with no tidy formula, which no one understands very well.

Calling these attractors "strange" is actually a declaration of ignorance. Whenever mathematicians use that name, it implies that what they are working with is important, even if they don't understand it yet!

That's not very clear at all! Can you give us an example of what you mean by a strange attractor?

Let's look at it in terms of the motion of a particle. Once a particle is attracted to a strange attractor, it can't escape. It will look as if it is moving randomly, but obviously the motion is specified by precise laws because, if you plot out the path followed by the particle point by point, over time you will have the eerie experience of seeing a shape gradually emerge like a ghost out of the mist. Slowly you'll begin to see the hidden structure of a system that otherwise seems patternless.

One of the first strange attractors was found by Edward Lorenz in 1963, and is called the Lorenz attractor. He was a mathematician turned meteorologist, studying climate change and weather. Classical types of attractor describe the motions of pendulums and planets. This new attractor is not a point or a circle, and is called "strange." As his powerful computer analyzed the very complex weather data, involving very complicated equations, a series of graphical curves appeared and a characteristic shape emerged on his screen. It looks like a mask with two eyeholes, or butterfly wings, but twisted so that the left and right-hand sides bend in different directions. It actually exists in an infinite-dimensional space (phase space).

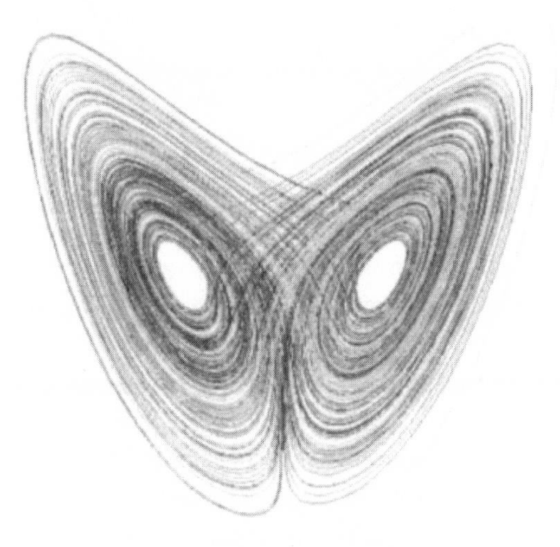

FIGURE 10.3 STRANGE ATTRACTOR

So where do fractals come into this?

If you tried to describe all of the patterns associated with strange attractors purely in mathematical terms, you'd get lost in the numbers. Fractals translate that mathematical description into a visual image. Moving computer-generated graphics make it possible for us to visualize phase spaces with an infinite number of dimensions. (We have been using *dimensions* in this chapter as a purely mathematical concept of phase space—perhaps *variables* would be as helpful here.) Sometimes the projection is frozen in one position, which gives us something like a cross section or a slice through the pattern. All of these visual representations help us to get a handle on this new kind of infinitely complex order, although obviously some information is lost in the translation process.

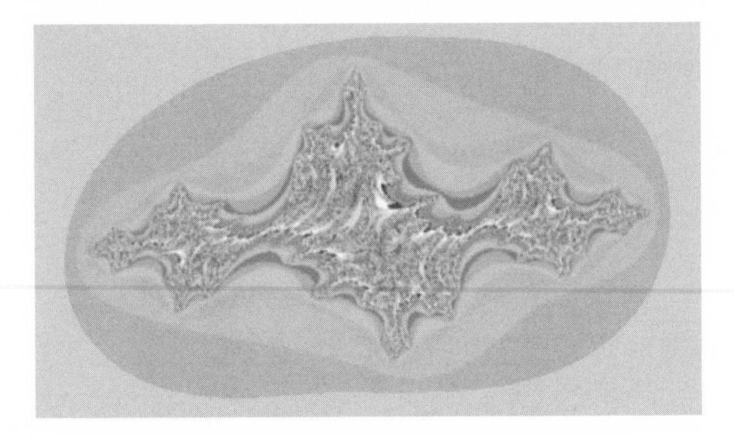

FIGURE 10.4 A FRACTAL FROM THE COMBINED BARNSLEY JULIA AND MANDELBROT SETS (MICHAEL BARNSLEY AND GASTON JULIA WERE ALSO MATHEMATICIANS)

You've mentioned some uses for fractal geometry in physics and geography. Are those the only areas they are used in?

No, they are also used in technology. For example, an understanding of chaotic signal distribution has been used to improve television reception. And the texture of many of the images you see on your computer screen is created with the help of fractals.

Biologists, too, have found that strange attractors provide a useful model for the controlled chaos of the brain, which is also characterized by "complex behavior that seems random but actually has some hidden order."[3] Walter Freeman has developed beautiful phase portraits made from brain EEGs (electroencephalograms). He argues that chaos theory explains the ability of the brain to respond flexibly to the outside world. Our alpha rhythms, for example, from the visual field of the brain, are constantly searching for patterns, the visual order within the apparent chaos of shapes that we perceive. Freeman thinks that consciousness may be the subjective experience of these processes. The brain also generates new activity patterns and fresh creative ideas to handle new situations.

Physicists are also trying to work out whether the quantum indeterminacy of the wave function, which we discussed in chapter 2, could be a manifestation of chaos.

Speaking about quantum physics reminds me of all those extra dimensions we were talking about then. How do these tie in with chaos theory?

The fascinating thing about these multiple variables is that when we do the mathematics using ten or more spatial dimensions, chaos disappears! It is replaced with patterns described in ten dimensions.[4] So superstrings (or even supermembranes) in ten or eleven dimensions seem to fit perfectly well with chaos theory too, which is another argument in support of their existence.

What are people who are interested in chaos theory studying now? I mean, those people who want to work at the edge of things, and not just apply what we already know about chaos to things like computer graphics.

One of the things that is interesting researchers now is the balance between order and chaos. Everything that is real is chaotic, but there is also a sense in which we live "at the edge of chaos," in a small pool of seeming order. And so an even newer science of "complexity" has begun to appear, which focuses on studying the interrelationships between systems in which many independent agents are interacting with each other in a great many ways. This new area is sometimes called chaotics or modern dynamics.[5]

You mentioned the dimensions we discussed earlier. But I have been thinking about something else we talked about, the anthropic principle. If fractals reveal patterns we would never

have suspected on our own, isn't something similar going on with the anthropic principle?

Yes. The anthropic principle says that if things had been ever so slightly different, life would not have been possible. The cosmos looks like a random mix of exploding supernovas and gases and the like, and yet when we examine it in detail, everything seems to have been carefully prepared and kept ready for us to be here. It seems that within the apparent chaos of the cosmos there is indeed a beautiful order—"determinism within chaos," as Mandelbrot put it.[6]

Many modern physicists would agree with Paul Davies' conclusion in his book *Superforce*: "The universe must have a purpose—and that purpose includes us."[7] John Barrow also finds "evidence of supernatural design, uniquely prescribed by some higher internal logic" and admits, "we are unable to conclude anything further . . . without appealing to metaphysical or religious beliefs."[8]

You know, what has struck me is the way fractals are pictures that give us a way to visualize dimensions we can't see. Isn't that similar to the way we talked about what the Circle did in Flatland?

Yes. The Flatlanders couldn't for the life of them envisage a third dimension, so the Circle took a two-dimensional shape in an attempt to help them understand that there was a Sphere. Or, to put it more in the terms Mandelbrot might use, the complexity of many dimensions would have remained veiled in mathematics, without any apparent application to the real world, had its visual nature not been revealed.

Of course, when we create a visual representation of a fractal, we are probably losing some details about the extra dimensions that are there in the mathematics. We are reducing it all to a two-dimensional image. In the same way, when the Sphere took the shape of a Circle, it lost some of its spherical attributes—but it could still tell the Flatlanders a lot more about the Sphere than they could possibly have known on their own. To carry this over into Christian terms, you could even describe Jesus as a fractal image of God.

Isn't that going a bit far?

I don't think so. Paul says in so many words that Jesus is the image of the invisible God,[9] that is, a representation of God that we could actually see and touch. And not in the way you can see and touch an idol, but more like a fractal image that moves and turns on your computer screen. Jesus was active, and in some ways he was even like a strange attractor.

What do you mean "Jesus is a strange attractor"?

Apart from the surface meaning of those words in ordinary language (which you'd probably find to be true, too!), I mean that looking at his life and the background to it feels like watching a pattern build up slowly. You have all the bits of information floating around in the Bible and gradually lining up to form a pattern that was there from the beginning, but couldn't easily be seen in the "chaos" of events. You see the hints of the pattern building up gradually over the centuries. Moses said that God would send another great prophet;[10] eight hundred years later, Isaiah prophesied about the life and death of a Suffering Servant;[11] and 800 years later Jesus arrived, and suddenly the pattern was complete and we could recognize the shape of the attractor. Paul recognized the pattern and got all excited about it: saying that "all that is in heaven and all that is on earth is summed up in him."[12]

But did other people agree with Paul? Did they also think that Jesus really represented God?

Well, John uses a similar image when he talks about Jesus as "the Word" who was also God.[13] Somebody's word, their voice and what they say with it, is the way they communicate with someone else and express their own ideas. So to say that Jesus is the Word that became flesh and lived here with us[14] is also to imply that he communicates God to us and tells us about God and about how he sees things.

OK. So the people who wrote the Gospels thought that Jesus represented God, but why did they think this? What did he say that gave them the idea?

He put it pretty clearly when Philip, one of the first disciples, asked him to show them this Father God he kept talking about, and Jesus blew them away with his reply, "He who has seen me has seen the Father."[15]

But of course it is easy just to say that. Trouble was, Jesus backed it up with what he did (particularly by telling them that he wouldn't stay dead and then not staying dead). And then there was the "strange attractor" bit, the way all the prophecies fell into place around him. That backed it up too.

The disciples, who watched him pretty closely for a number of years, ended up convinced that no one else can tell us as much about what God is like and about how he feels toward us as Jesus can. Jesus alone perfectly reveals all that God is and will be.

So you're saying that God has communicated with us, and that Jesus is like a fractal image of God. And this stuff about strange attractors probably means that the chaotic mess we see in the world around us isn't really a mess, but just something orderly that we haven't figured out the order for yet. I'd buy that, except for the fact that so much of what happens is really BAD. If God is as good as you keep saying, and as Jesus represents him, how come there is so much suffering in the world?

That's a question that has puzzled theologians and philosophers and scientists for centuries. I think we need another session to even begin to discuss it.

11. But Why? The Problem of Evil

The Spirit brings order out of chaos. —JACK DEERE[1]

THE EVENTS OF 9/11; 7/7 in London; the bombing of the World Trade Center; the taping of the screams of tortured children by Ian Brady and Mira Hindley; the massacres in a primary school at Dunblane, Columbine High School in Colorado, and at Virginia Tech; child abuse; Hiroshima; Auschwitz; apartheid; ethnic cleansing; child soldiers; massacres in Rwanda and Kosovo; the Israel/Palestinian conflict; the Iraq War; and, until recently, the conflict in Northern Ireland—how are these to be understood? They provide abundant evidence of the chaos in this world. But I had been saying that apparent chaos was often the working out of a hidden order.

How can you possibly say that there is any sort of order in a world where this type of thing can happen? And you even implied that God created that order. I'm not interested in a God who acts like that!

Before we start looking at God, let's look at ourselves for a bit. Have you heard about the studies in obedience that were done a few years ago? Researchers found that ordinary people, often students, were quite prepared to inflict severe pain on others if they were given clear and seemingly logical instructions to follow. Very, very few of the people who took part in the experiments refused to hurt someone.[2]

Those experiments were generally carried out in the calm environment of a laboratory. Do you think ordinary people—you, for example

–would fare much better in real life? Particularly if your emotions were strongly involved.

I like to think I would be one of the people who refused to participate. But realistically—I might have gone along too. And even if I had resisted in the laboratory, in real life there have been times when I've been so angry that I could have done something really awful if I'd had the opportunity. But what's that got to do with the first question we asked?

We like to think that we're too cultured and sophisticated to do anything evil. But those experiments, and our own experience, show that some dark forces are very close to the surface in many of us. But what's interesting about this is that we don't seem to see any similar patterns of evil in animals. Even if 98 percent of our genetic material is identical with that of chimpanzees, chimpanzees don't conduct all-out wars or torture other chimps for the fun of it. Sure, they squabble a bit, but they don't go to the lengths that humans do. Once again, it's clear that humans are different from animals in some fundamental way.

Last time you talked about why humans are different, you said it was because God intervened in some way and gave humans what you could call a spirit, or a soul. But surely if God is good, he wouldn't have given them something that made them cruel.

Yes, and that is exactly why the story in Genesis doesn't stop with the Creation, but carries on to give an inspired account of the true origin of evil.

Adam and Eve represent the first human beings to have consciousness, an awareness of God. But once we are consciously aware of something, we have to decide what to do about it. If your radio is on so loud that you don't hear your mother calling you for dinner, you carry on quite happily doing whatever you were doing. But when you become aware that she is calling you, you have a choice: either carry on—in which case you are ignoring her—or arrive at the table. In a similar way, once Adam and Eve knew that God was there and had an interest in how they lived, they had to choose how to respond.

What they chose was to ignore him, to go their *own* way—and God let them do it![3]

But why would he do that? He could easily have designed things so that they would always live the way he wanted!

We discussed that when we talked about Flatland. Could the Sphere have had a real relationship with the Flatlanders if he had dictated all the terms? It would be a bit like having a relationship with a computer program. Or he could have put on such a dazzling display of power that the Flatlanders could have been forced to love him or at least respect him. But we know that even in human terms that doesn't work. No man can compel a girl to fall in love with him.

For some reason we cannot fathom, the all-powerful God wanted a real relationship with some of the creatures he had made. But in doing this, God created the possibility that these beings would choose the exact opposite to what he wants. Love is more important to God than a perfect world. He is willing to take the risk of disorder, rebellion, and suffering, even pain and death in the creation, if only there can be love!

So God is content to work to achieve his purposes through the patient unfolding of a process. "Perhaps there is no other way for love to work if it is to respect the integrity of the beloved."[4]

Why was the decision to "go into business for themselves" and ignore God such a bad one? What were the consequences?

In some ways, ignoring God is a bit like ignoring an old, wise friend. That friend may not reject you in response. He or she may still want to be friends with you. But if you don't take the time to talk, you lose the benefit of the advice that person could give you about decisions that affect your life. And if you do see that friend in the street, you may feel guilty that you haven't contacted her, and cross to the other side to avoid her, or get angry with him when he asks where you have been and how you are doing: "It's my life. I don't have to report to you!"

When Adam and Eve chose to ignore God, their friendship with him followed this path. Consequently they ended up doing more and more things the wrong way, and guilt and anxiety led to fear: "And so we hid ourselves from the Lord."[5] The book of Genesis isn't about science, but it does give us a vivid picture of the true origin of sin, guilt, wrong judgments, and evil.

But what I have said so far isn't the whole story, of course. Because Adam and Eve didn't just reject God as a friend, they chose someone else as their friend and adviser, and "got into bad company" you might say.

Come off it! Surely you don't believe in a real devil, with horns and a tail and a pitchfork, who disguised himself as a snake to tempt Adam and Eve.

You know, you get your pictures of Satan from the same source as you got your picture of heaven—comic strips! A really reliable source of information!

In the Bible, Satan is a lot more subtle than the popular image of him. His goal is to disrupt the communication between Creator and creation and to do this he uses tools like slander and accusation. He slanders God by telling us God's unfair and unreasonable; he accuses us of things like having the wrong motives for whatever we do, or of being such failures that no one, not even God, has any interest in us. In fact the very name *Satan* actually means "the accuser of the brethren" or "the opponent." The person he is primarily opposed to is God, but you and I are pawns in a bigger battle for control.

When we talked about Flatland, we didn't even think of a character like this. Where does he come from? What's his relationship to the Sphere?

I never pretended that our Flatland analogy told the whole story—analogies seldom do. But one of the points that we did make was that the Sphere had a very hard time communicating with the Flatlanders, and maybe that wasn't just because of the different number of dimensions.

But let's focus on the different dimensions for a bit. Science has confirmed what the Bible already told us, that there is a lot more to life than meets the eye. This "more" includes a spiritual dimension that we ignore at our peril.

We don't fully understand the relationship between the Creator and the Adversary, although we do know that the Creator is more powerful.

But because he gave us the freedom to choose our friends, we were able to choose his enemy as our friend. And that friend brought with him a flood of other evil beings who sow misery in this world and keep it in spiritual darkness under the domination of the Adversary.

But why would God let his enemy control what he had made, particularly if he is more powerful?

We keep coming back to the fact that God gave human beings a free choice. And God isn't like some politicians who love democracy until a vote goes against them! God sticks to his choices. He has limited his own freedom rather than contradict his gift of free will. Bad things happen, not necessarily because God wants or wills them, but because he cannot prevent them without taking away freedom. In some ways, this is like parents having to let their children make real decisions. Parents teach them values, warn them of dangers, spell out consequences, but the child is ultimately the one who has to decide what to do. Parents cannot make all the decisions for their kids if they ever want them to become adults.

Does this mean that God just lets evil have a free hand in the world? Doesn't he do anything to limit it?

In a sense, God practices tough love: humans have made their choice and they have to live with the consequences. Of course, one side effect of this is to show us how bad evil really is, and to make us more aware of what we were rejecting when we turned our backs on God. Even the psychologist Carl Jung recognized this when he argued that if evil is not allowed to exist, then good itself would appear shadowy.[6] We also have a real example of this in the experience of General Dallaire, the head of the UN forces in Rwanda, who desperately tried to avert the massacres there, but whose calls for assistance were ignored. He says that he is more convinced than ever that God exists because he has "been to hell and seen who the devil is."[7]

But God doesn't just let us suffer the consequences; he also works to turn evil round, to use it to produce good. For example, he allowed Jesus to be betrayed and crucified unfairly, but that evil produced the ultimate proof that God cares and wants to connect with us again.

Sometimes we see him doing this type of thing in the lives of people we know. But it is more difficult for him to do this in our own lives if we don't allow him to do it—there's that issue of freedom of choice again!

I want to get back to what you said about the Devil, or the Adversary as you preferred to call him. Do you really believe that there is actually such a thing as an evil spirit? I can handle talking about natural forces, or even the "subconscious" as guiding our actions, but I find it awfully difficult to accept that there are actual evil beings of some kind.

So do I. Our culture tends to dismiss any idea of angels, demons, occult powers, and the like as fodder for Hollywood movies, and nothing to do with real life. But most of the world, except for Europe and North America, takes the existence of evil spirits for granted. They recognize the reality of the dark powers and the need to confront them.

But we used to. And we have had a long struggle to get away from conducting witch hunts and other awful things. Do you want us to go back to that type of life?

The trouble is that modern secularism has been so anxious to get rid of every trace of transcendence, of a world that exists beyond material things we can see or touch, that there is massive resistance to even thinking about these phenomena. We deal in caricatures, often in themselves evil, like the witch hunts, but not with the things themselves, and we have ended without any language we can use to discuss these things.

But if we dismiss all spiritual forces, and don't talk about them, we also end up unable to talk deeply about who we are, or about who God is. We can't just talk about those parts of reality that we like.

You've said that quantum physics and the anthropic principle are undermining the materialistic approach that ignores other dimensions. But so is all the New Age stuff. Some of it seems to involve a lot of contact with spirits and things like that.

You're right. Materialism hasn't satisfied, and many people sense an acute spiritual hunger and are reaching out for spirituality. Sometimes they search for this within themselves, which is not a completely good idea because, as we discussed earlier, darkness and hurt can lurk pretty deep inside us all of us and lead us to feel all kinds of actions are justifiable.

People who are aware of their own weaknesses sometimes prefer to make contact with other spirits through channelers, mediums, Ouija boards, Tarot cards, Wicca, and the like. But they often forget that spirits are not neutral, but can be good or bad, and so people are actually often linking up with spirits of darkness and superstition, occult forces. The Adversary wins a round each time people prefer to hang out with these spirits rather than listen to God's Holy Spirit.

OK. But I presume you are supposed to go to church to listen to God's Spirit, and church takes so long and is so boring. Now if I go to a rave and take Ecstasy, or attend a séance, that's exciting. If God's Spirit is better, shouldn't it also feel better?

That's the type of question to which the answer is both yes and no.

Yes, church can be pretty boring, particularly when it is merely a lifeless "religion" with rituals, traditions, doctrines, calendars of saints, and rules, but without any awareness of the God who is behind these things. That was the type of religion rejected by Don Cupitt, Richard Dawkins, and others. I found it pretty deadly myself.

No, the good isn't always more exciting than the bad, just like junk food doesn't taste worse than healthy food. The way to judge something isn't what it does for me now, at this exact moment, but what it leaves me feeling like in the long term. A diet of hamburgers and chips tastes great for a few years—but the excess weight and the health problems you eventually run into aren't nearly such fun to deal with. It's the same with smoking: it feels good now, but less so when you get hit with lung cancer or emphysema.

I found church and Christians pretty boring too. (I have the distinction of having been expelled from Sunday school.) But all that changed, and now my experience is that the long-term effects of following God's Spirit more than make up for missing some of the short-term excitement. Plus, of course, I have run into all kinds of excitement I would never have expected. God's Spirit has a way of doing that to one.

Tell us about it. What changed your mind about Christians?
It wasn't so much my views of Christians that changed, as my view of Jesus. I drifted around a couple of churches as a teenager, but what was going on didn't make much sense to me. Then I went on to university, and one day found myself listening to what I expected to be a rather dry talk by a law professor.[8] It grabbed me! For the first time in my life, I heard that Jesus died for me, to give me a way to establish contact with God. I discovered that, in sending Jesus, God offered everything of himself, not to remove evil, but to come into people's hearts to help them overcome it, if they would accept him. It hit me how much Jesus loved me. That evening, I gave my life to him, to be used in his service. The following morning I experienced the joy of his presence as I read my Bible with new eyes, new understanding. I found the power of the Bible as a creative word through which Jesus spoke to me. And it still does, more than thirty years later!

You also said that you had some unexpected excitement. Tell us about that.
Well, the feeling I had when I accepted Jesus also involved a good measure of unexpected joy and excitement, so I wouldn't dismiss that too casually! But you're right, there was more to come.

One day Andy, a colleague, asked me to talk to a group of atheist and agnostic students, rather like this group, about science and belief. I tried to show them that perhaps science hadn't got all the answers, and our discussion got quite heated, but we didn't really seem to be getting anywhere at first. Then gradually the atmosphere changed until they were becoming very open and asking searching spiritual questions. I was amazed: my answers hadn't been that brilliant. What had made the difference? It was then that I noticed that Andy was not taking much part in the discussion, but was sitting in the corner praying quietly, and God was responding with a minor miracle that involved me! I was stunned.

So prayer really does work?
Well, it seemed to me that Andy had some kind of "hot line" to God. I wanted one too. Andy then told me about meeting the Holy Spirit and about the power of the Spirit to change things (with our help!), in this

actual physical world. I asked for the gift of the Holy Spirit, too. Nothing happened immediately, because God had some things he needed to work on in me first, but a month or two later I found myself with eight hours spare just to listen to God—and to his Spirit.

Did you go on some sort of meditation or fast or something like that?

No actually, I did something really spiritual: I flew from England to the U.S. to attend a conference where I was to present a paper. I had a book with me on the healing of memories,[9] which I wanted to read because I was still not at peace with my father about things that had happened many years before, and upset that I had not been able to say goodbye to him properly. I opened the book, praying that I would be able to hear whatever the Spirit might want to say to me. What the Spirit did was remind me of how Jesus was with me at the time when my father was dying, and of Jesus' love. Suddenly, I felt enabled to forgive my father, and also to recognize where my own actions and attitudes to him had been wrong. My pride in my own achievements and my academic degrees crumbled—even my pride in the paper I was going to present. I realized that knowing Jesus and having his forgiveness was more important than any of these other things.

It was as if an enormous burden lifted. The Spirit gave me the gift of repentance and of tears (the latter a surprising gift, given that as an ex-boxer for Cambridge University I am not in the habit of crying) and an almost physical, tangible awareness of his presence. I felt completely renewed and healed. From that moment I became conscious that I could invite the Holy Spirit to be present at any time. Now I understood what Paul Cho meant when he said that "the anointing of the Holy Spirit that comes through prayer . . . belongs to a higher dimension than just natural wisdom and understanding."[10]

And it's not just in me that I've seen the Spirit work—I have seen him working in others too, and that is really exciting.

Does he ever work in groups like this one?

Yes, but I'll only tell you about one of the folk he touched. It was a very artistic student with intense blue eyes, black eyebrows, and ruffled hair called Josh. Josh had fooled around with spirit guides and was very into heavy metal rock music, and had accepted a friend's invitations to a spiritualist "church." After attending this, he became aware of very powerful dark spirits moving within him, and was very disturbed about it. Eventually he came to me in my rooms at college and shared his distress with me. I didn't know what to do, so I simply listened to everything, while silently calling on Jesus to help. Josh sank into a deep silence, too, and then he suddenly shook himself, as if waking up: "It's gone—I'm free!" He was filled with joy and amazement. He had felt paralyzed, his legs heavy, unable to get up or talk, but somehow the power of Jesus had broken through and released him from the darkness, which never returned.

Ever since then the two of us have shared a special joy that the Holy Spirit is always present with us, and that the risen Jesus is more powerful than any demonic intruder. It's a constant reminder to both of us to give praise to the Father in every circumstance. And Josh has used his talents to present his experiences very vividly to other students who are tempted to doubt that dark spirits exist and can attack us.

Interestingly, Josh found that the Spirit of Jesus could also use him to free others from powers of darkness. But he had to learn not to go too fast, and never to work on his own without fellowship and prayer support. He has also found that unless the person he prays for really accepts Jesus as Lord, some of the darkness will return. He has become very wary of going ahead of where the Holy Spirit leads.

That's really weird! It's almost as if we are living in two worlds at once: the one we can see and one that we can't.

The Bible clearly teaches, and Josh's and my experience confirmed it, that we live in two worlds. There is a natural, material world as well as an invisible and spiritual world. And in both worlds, there is both good and evil.

Many people are quite happy to believe in a supernatural God because it's comforting to think that he is there in case of emergency. They are a lot less happy with talk about the Holy Spirit or about the devil or demons. We tend to want God only to act when we ask him to, and do what we want. Having the Spirit of God around and liable to act powerfully with a tangible presence, like breath or warmth, can be worrying to us. We like to know that we are in control.

Yet all the time people are having experiences of transcendence that challenge us to be aware of the Spirit of God. For some it is a wonderful experience like light, or a sunset for Josh, joyous music for Ruth, surfing or rock climbing for Angus, or the joy of a newborn baby for my colleague, Andy.

All the examples you are giving show spiritual forces operating in individuals. Do they also work on a bigger scale?

That's a bit of an odd question to ask when you look at the world around us! We certainly see evil forces at work in social institutions and even nations. We see them when society accepts that the highest goal of a corporation is to make money, regardless of who or what gets hurt in the process. We see them when we accept the assumption that only what can be measured and quantified is important, and that everything else can be ignored. We see them when we accept the use of nuclear warheads, cruise missiles, napalm bombs, landmines, and other weapons.

And we need to watch out for them among ourselves. After the Second World War, one German pastor stated, "You cannot understand what has happened in Germany unless you understand that we were possessed by demonic powers. I do not say this to excuse ourselves, because we let ourselves be possessed."[11] Remember that the Adversary is subtle. He likes to get leaders to take his advice—think of the suffering imposed by evil men such as Hitler, Pol Pot, and Stalin. But what the Adversary really likes to do is to make the use of evil instruments and the pursuit of evil goals look like common sense, the only possible thing to do.[12]

It is Jesus who reminds us of the other possibilities, that God and others are more important than profit and that we need to love even our enemies.

That's an ideal that sounds great, but is really difficult to put into practice. Is it even possible?

Of course it's difficult—we have a very strong opponent suggesting the opposite. But that is why God has given us his Spirit, to help us to fight the Adversary, who otherwise would be far stronger than we are.

With the Spirit's help, our bouts with evil forces, our temptations, don't drive us away from the Lord, but toward him. God then uses the temptations as opportunities for self-discovery and pruning, to shape us into the people he wants us to be.

Yet sometimes the evil we meet is so overwhelming that all we can do is take Jesus' advice in the Lord's Prayer and pray to be delivered from the encounter. Real evil is still a mystery, yet our awareness of the powers of darkness also motivates us to rely on the Holy Spirit to overcome the Evil One. Josh and I found that what matters is that we encounter the risen Jesus in the power of the Holy Spirit, and learn to walk with him.

If you are tempted to be skeptical about the reality of the two worlds, remember that the ultimate scientific question is "Does it work?" My experience, and Josh's, and that of many people I know and many others I do not know is that it does. Just as our science led us into a transcendent reality beyond the physical three dimensions—into black holes and singularities, superstrings in ten dimensions, and the hidden order within multidimensional chaos—so our spirituality inevitably leads us beyond the physical into a deeper reality. We need to recover a full-blown understanding of reality as including a spiritual dimension.

12. The View from Here

These are a shadow of the things that were to come; the reality, however, is found in Jesus. —PAUL[1]

THE TERM was coming to an end, and the students would soon be scattering to their homes, or to new jobs. Our weekly sessions would have to come to an end. But before we separated, we wanted to have one last meeting to review what we had been studying and to give everyone a chance to ask their final questions. I flung in the first question myself.

Do you remember what you said when we first discussed this group? You, Richard, were adamant that you only wanted to talk about science, not religion. But you also wanted us to try to develop a "theory of everything" that would be big enough to embrace your science and Melissa's spirituality or (as I think you hoped) would dismiss Melissa's views and support your view that science has all the answers. Did we manage?

Well, the first thing I discovered was that my view of science was hopelessly inadequate. I could handle the size of the universe, as long as we kept it to one universe. But once we started getting into quantum physics, and M-theory, and all those other dimensions . . . Each time scientists solved one problem, they seemed to come up with a new question, and each question was more bizarre than the last. It certainly didn't give me the secure sense of explaining everything that I had wanted.

So, did you come around to Melissa's views, that there is more to life than science?

I suppose so . . . although not the way she said it at first, when she seemed to imply that science had nothing to say about the spiritual world. The way you explained things, there does seem to be some overlap. I suppose it's a bit like that duality you talked about. Things that look very different, like steam and ice, a wave and a particle, or different forces, can actually be the same when you look at them from the right angle, or, from what you've said, with the right number of dimensions. So it seems that it is possible to be a serious scientist and to explore the possibility that God exists. I'd never have believed that before.

My parents certainly won't believe me if I tell them that what you said about the anthropic principle is making me rethink all that antireligion stuff they've been teaching me since I was a kid. It's scary . . . and a bit exciting.

Why scary?

Scary, because, well—without a Creator, there was no chance that my life had any particular purpose, so I could just relax and enjoy myself: "eat, drink and be merry" and try to forget about the "tomorrow we die" part of the proverb. Or even if I did think about it, it was more with a sense of frustration that I'd just stop existing when I died, and all my ideas and the people I loved would also just evaporate, as if we had never been.

But now—if there is a Creator, and you have pretty much convinced me there is, he seems to be pulling strings sometimes to make things happen. I don't want anybody else organizing my life! And yet it's stupid to say that, because even as an atheist I know that what other people do affects me in ways I have no control over. Someone drives drunk, and I am gone, regardless of what I do. I suppose it's better to have a mind directing the world than just chance, but a mind is scarier because it has plans that I don't know about.

And also, I need to decide what to do about that Creator. You said he doesn't force us to respond to him. So do I just carry on as if he's not there? Or do I try and work out why he put me here and try and live in a way that fits with the design specs? And what will happen when I die—

you implied that if I ignore the Creator, he'll just ignore me back. But what a waste, to have the opportunity of having one's life carry on in a different dimension and throw it away because I couldn't be bothered to listen to God!

You've made me more scared because now the decisions I make have a lot more implications than I ever thought. But I suppose on balance it's better to know the score rather than to carry on being ignorant.

You're right there. Were there other things that some of you found frightening?

What made me nervous was when you suggested it was a bit stupid to assume that the spiritual world was automatically good. I've fooled around with druids and Ouija boards and Tarot cards and things like that, and I didn't like the idea that I might have been talking to dark forces, and giving them an entry into my life. So I went out and bought a book about Tarot cards written by a Christian,[2] to see what he had to say. He also felt that they could open a door to evil forces, but I was amazed that he kept on explaining the Christian ideas in the pictures on the cards. He was doing the same thing you are doing—showing me that God is hiding out to ambush me in unexpected places! Angus thought science was safe from Christianity; I thought that if I stuck with the Tarot, I'd never get to be a boring Christian. But like you said, the Spirit takes apparent chaos and random things and directs our thinking in ways we couldn't have imagined. I don't find it quite as scary as Angus does—more exhilarating!

So what do you think the Spirit is saying to you?

So far, I think I've only got part of the message. I've heard enough to convince me that God exists, and that the world has dimensions we know nothing about, some of which are really nasty. I keep thinking of myself as one of those Flatlanders. I know there's another dimension out there, the Creator's dimension, but how am I going to communicate with it?

Doesn't the Flatlander analogy help you there too?

How? Wait . . . in Flatland, it wasn't the Flatlanders who set up the communication: it was the Sphere who initiated the whole deal, and went to a lot of trouble about it . . . I suppose you are saying that Jesus was like the Circle, God's most powerful communicator with Flatland, and the one who made real communication possible, and proved it by the Resurrection. But he isn't here any more, or at least, not here in the same dimension I am.

But isn't that where his Spirit comes in?
I suppose so. But I am still puzzled. When I talk about someone's spirit, I mean their general attitude to life, and maybe a sort of ghostly presence that hangs around after they're dead. But you keep speaking as if the Spirit is a lot more than that. Almost like a person who can communicate with us.

Exactly. I'm glad you noticed. Have you ever heard Christians talk about the Trinity? What do you think they mean?
I've never thought about it much. But I suppose what they mean is that there is God, the Creator you've talked about in the science. And then Christians also say that Jesus is God (although he was also human, a Flatlander like us), so I suppose the Spirit must be the third part. But it sounds really weird to say that three "beings," for want of a better word, make up God.

Is it much weirder than what we've been learning about in quantum physics?
No, I suppose not. I keep having to shake myself and remind myself that reality isn't at all what it seems! There are different dimensions out there! And I suppose, from M-theory, those different dimensions aren't just out there but also right here, curled up alongside me.

. . . . .

But by this time the students were tired of my questions, and again brimming over with their questions for me.

It's all very well for you to talk about the Trinity and all these things, and about Jesus coming to set up communication between us and God. But in practice, when I look around, I don't see many effects of his coming, and certainly all the sentimental gush about "peace on earth" at Christmas is false. Does Christianity really work in practice?

I've two responses to that. First, remember that what you see is just the smallest snippet of a very long history. Lots of the things we take for granted today (like democracy) are actually the product of years of Christian thinking—just as some of the other things we see today (like racism) are products of years of non-Christian thinking, some of it unfortunately done by Christians too!

But the second point relates to what is actually meant by "peace on earth," which often isn't anything like what the Christmas cards suggest. That "peace" isn't referring primarily to everybody stopping fighting here. It's got more to do with the spiritual wars in which we are the pawns.

Remember that I said that Adam and Eve chose to avoid God and link up with other supernatural forces. At first, the link didn't seem too bad, but people gradually slipped more and more into small individual sins, like cheating, lying, betrayals, and selfishness. (Although these are only "small" compared to the original sin, which was deciding to ignore God.) We feel sins like these as a pretty heavy burden, but still they're a burden that we quite like. After all, it's convenient to be able to lie to get out of trouble. Though we're not quite so happy about it when someone lies to us . . . or worse. And our consciences do bother us, unless we can drown them out.

This piling up of sins, and the fact that we had stopped speaking to God so long ago that we had lost touch with the spiritual world and forgotten where to find him, meant that the initiative in solving the problem had to come from God. And the peace he offers is the chance to make peace with him, instead of siding with his opponents.

So how do we get in on this peace treaty?

By listening to what God has to say in the Bible, which lays out his action plan. I can summarize it very briefly as involving our telling God that we recognize why Jesus came, are deeply disturbed about the fact that we have been on the wrong side, and about the things that are wrong in our lives, and asking to share in the benefits of what he has done.

Not everyone is prepared to do that. I mean, it's humiliating to admit you've been wrong, and it's pretty hard to change a lifestyle you've lived with and been comfortable with for years.

But even the Christians who have accepted what Jesus did don't always seem to live what I would think of as a life in touch with God. They still sin. And some of them are still hurting and unsure that God really loves them. If Jesus conquered death and sin and all that lot, why doesn't he fix them up immediately?

Jesus didn't promise that life would be a picnic for Christians—in fact he promised the opposite. "Put on the full armor of God, (the shield of faith, the sword of the Spirit which is the word of God. . . so that you can take your stand against the powers of this dark world and against the spiritual forces of evil . . .) And pray in the Spirit on all occasions."[3] He doesn't whip them out of this world into the spiritual dimension. Think about it: most of this world is the Adversary's territory, and those people had been on the Adversary's side before. Now they have switched allegiance. The Adversary is going to use every trick he knows to make them act badly, look like fools, or keep them from really feeling forgiven and conscious of the Holy Spirit.[4]

Jesus knew this, which is why he sent his Spirit to help us stand up to this battering. Sure, we fall down every now and again. Sometimes external things trip us up, and sometimes it's internal things, our own memories and emotions; but the Spirit comes alongside if we ask him to, helps us back on our feet (or props us up so we don't fall) and says, "You're on the right track—keep heading into that new dimension." And every now and again he gives us a glimpse of where we are heading, and of the fact that for us death is a doorway into a new and exciting life, a chance to explore life in a new dimension.

It sounds great! But what do I do now?

Keep listening. And that applies to all of us, whether we have accepted Jesus or not.

Listen to what the Creator says to you in the universe. Cosmology and the anthropic principle tell you about his awesome power, which balanced immense forces so precisely to allow life to emerge. Quantum physics and string theory open our eyes to the fact that reality is infinitely more varied than we could ever have imagined. There is no reason that God cannot be beside us and within us without our knowledge. Chaos theory tells you that apparent disorder may conceal a most profound plan.

Listen to what the God of love says through his Word about his attempts to communicate with us, and about our refusal to listen.

And when you have listened, answer. If the Creator/Father speaks to you, don't hold back your allegiance for fear of the ridicule of your companions, who are still mesmerized by the two-dimensional shadows in the cave. Escape into the glorious freedom of the sons and daughters of God.

13. M-theory in Eleven Dimensions: The Spiritual Universe

"I'm going to give you two endings" was my cryptic response to an earlier request for a wider "theory of everything." This chapter is more philosophical and holistic, even prophetic, using these current scientific ideas to resonate with our search for a deeper reality. The other was more personal, encountering dimensions of the spirit—where real prayer changes things. "It works," as recent clinical experiments seem to reveal.

The challenge was to make sense of a spiritual reality beyond the reductionism of the three-dimensional world picture of Richard Dawkins, who focuses on the physical, ruling out any dimensions of the spirit. We shall move beyond the positivism of Stephen Hawking where only what can be tested and proved is true, to the models of transcendence in today's physics.

We wanted to make sense of the mysteries and paradoxes in science, as well as being able to find a language for talking about prayer, miracles, healing, the Resurrection, and eternal life in the context of a multidimensional worldview.

My prophetic understanding is that the new scientific world picture will become a multidimensional worldview to enhance our perception of a holistic reality.

"Can you make sense of heaven and eternity in the light of the current theory of everything?" queried Angus. "We've been on an exciting journey of discovery—this might be the final challenge—to link it with our spiritual discoveries," Richard, the down-to-earth scientist, stated. Where to start!

"What about that new M-theory in eleven or twelve dimensions that you were excited about?" Josh was into multimedia in his art studies, keen to translate the indescribable into visual metaphors. Melissa was still vitally interested in the spirit world. As a budding psychologist, she wanted to put to rest some of the dark powers that still plagued her. Richard, however, wanted to begin with the science: "It does seem as though M-theory is a fantastic underlying theory of everything." "And still can't be experimentally tested," teased Angus. "I've more proof of the presence of the Spirit!" laughed Josh. The others sided with Richard: "We think we know where we are, even with today's science." However with this new theory of the twenty-first century still catching everyone's imagination, we will have to accept some mystery with it—for it was M for mystery or magic, or the mother of all theories.

Melissa remembered the strange ideas of the 1980s, of particles not being points, but tiny loops or strings, which vibrated like Ruth's violin strings. "A cosmic string quintet," joked Ruth, "or a well-tempered universe!" This was to lead to superstrings in ten or eleven dimensions.

And then they found not one but five different superstring theories as well as supergravity in ten or eleven, Richard pointed out, though he doubted whether it was real: "How can all these be one grand theory of everything?"

Superstrings and even supergravity are now identical in one unique framework using the creative concept of duality. "A very exciting prospect," Josh enthused. He already saw enormous implications for talking about God, Jesus, and the Spirit. He wanted to press on; an idea was forming in his mind, steeped in multimedia pictures, of duality and the Trinity, of Jesus as Son of God . . . but the others wanted to focus on the science.

In the big bang, our customary three space and one time dimensions had expanded. The other seven space dimensions had contracted. "Until recently," Angus reminded us, "we thought these seven would be all compactified—curled up very small." Richard confirmed from Brian Greene's book[1] that one or more of these extra dimensions may well be comparatively large or even infinite.

"Therefore they could be tested in the next generation of particle accelerators," Ruth enthused—an exciting thought, Richard agreed. He had explored further. In Stephen Hawking's *The Universe in a Nutshell,* current M-theory suggests that there are not only strings (in one dimension or "brane"), but vibrating two-dimensional membranes (2-branes), and even three-dimensional blobs and a host of others—"p-branes." Their inventor, Paul Townsend, calls them "a democracy of branes"![2]

Josh and Melissa were enormously excited: "Just a fraction or so away from us at each point in spacetime is a shadow world." Richard was also entering the thrill of the visible universe like a 3-D brane, floating within higher dimensional space, "So we are only shadows (branes!) of a deeper reality, a higher dimensional spacetime"—"shades of Plato's cave," interrupted Ruth. "So it would provide a super model for your faith-stuff!" Melissa accepted this new world picture.

The spiritual dimensions

And so the group began to see a deeper pattern emerging from M-theory for the Holy Spirit and healing, as well as the dark shadows of ley lines, white witches, and Wiccas. There is certainly a widespread awareness of the supernatural. As Josh knew, there were also the dark powers—a lower dimension, he thought—the occult dark dimension that he had met before. Josh knew these dimensions also included the numinous feeling of the stone circle, which had triggered the whole experience of the Flatlanders many months ago.

Out of the cave—a look at eternity

Then a visitor asked if he could join us. Phil had heard about M-theory and wanted to know how this fitted in with the idea of eternity. He felt that these multidimensional theories had a deeper significance. We weren't sure we could answer this. A unified understanding in theoretical physics involves new concepts, "extra dimensions beyond the physical three," as Martin Rees has pointed out.[3] The Astronomer Royal also warned that previously held intuitions about space and time "will have to be jettisoned." This confirmed our confidence that a new world picture was being more widely accepted.

We had shared some out-of-the-cave experiences. "Maybe we thought we lived in a three-dimensional world, but are really shadows cast by the higher dimensions revealed in M-theory," pondered Angus.

"Could this be an analogue of the Incarnation?" wondered Richard, rather doubtfully. As a positivist, he couldn't decide what reality is. He realized he may have to make a leap of faith, perhaps "What if . . .?"—as Kaluza once tried.

We were all wary of "eternity" perhaps because of its being bound up with the concept of time: time for Richard was just a succession of moments, a clock ticking since the world began—until it ends. "This is the arrow of time," as Richard echoed what many scientists assume. Or as the atheist Peter Atkins might insist, the experience of time is merely the gearing of the electrochemical processes in our brain to this "purposeless drift into chaos" as we sink into equilibrium and the grave.

"You surely don't believe that," demanded Josh. Even Richard Dawkins' militant atheism (from his rather narrow biological standpoint) admits to a feeling of awe over living things.

"It seems to me," interjected the pragmatic Richard, "that there are two times. One is the 'chronos' dimension of science, the regular progression of seconds and minutes and hours. But there is also the 'kairos' dimension—the multidimensional moment that takes us out of the mundane, the everyday. Kairos is the creative, perhaps ecstatic time of love, of the spirit, seen in moments of transcendence."

We all remember beautiful sunsets over the hills, a baby's first cry—for some even the scoring of a vital goal . . . But there is also the dragging time, even boredom, and the torturer's time (the apparently everlasting time in solitary confinement, which Richard Wurmbrandt once described to us). "Indeed of Jesus on the cross," mused Josh.

Richard reminded us that time (and space) technically begins at the big bang. At that point all dimensions are unified at so-called t=0. If there is a big crunch, chronological time will cease then.

Even here the only point that we can conceive is our time, starting from the self-awareness beginning in our childhood, up to "now," our human limit. "Eternity was beyond us, literally," joked Angus. The Catholic idea of God in everlasting time or space seemed to have the wrong concepts, an inappropriate "category error." Josh preferred to think of God being involved with his creation, changing and interacting with his universe. "Needing us," as the process theologians might say.[4]

Ruth, our musician, had been particularly interested in healing time, and in letting ourselves into the dimension of music.

She preferred a string quartet where her violin sang in another dimension. "We haven't begun to think about time unless we talk about music," she said. "It's like contemplation if it goes well. Otherwise it becomes repetitive and boring." As Angus remarked, "I think that's what happened when I used to go to church services." "When the Spirit is present, it's so different," Josh reminded us. "The hard question: is your music linked to the timelessness of eternity, a sort of taking time out?" Josh thought we only really exist when we are aware of the higher dimensions, "a kind of redeeming the time?"

God is outside time—but also within it

Eternity, then, seems to be real only when it is now! The past does not exist any more. The future has not yet appeared! Josh preferred the more personal "everlasting life": "it is both now and not yet."—*now* in the awareness of the Spirit, and *not yet* as regards to the fulfillment of everything in the end times.

"I think God's eternity is not endless time," Richard was thinking on his feet. "He is not on some kind of endless time line, and he has allotted us all the time we need to fulfill our potential—our destiny if you like."

"Music is like that," mused Ruth. "Alternating sound and silence can be magic, another dimension of reality." "Sort of working with the grain," was how Richard looked at it. "Or patterns of equilibrium and tension resolving by the end," Melissa put in. "It's like suspending resolution until the promise is fulfilled."

"Have you heard any of John Tavener's music?" I challenged them. It seemed to me that music itself was a true redemptive analogy of the Spirit. Tavener believes his music may well be tapping into "a deep longing for the sacred"[5] in our lives. His critics say it's just ritual repetition, rather like substitute religion. "I'm happy with Mozart, or Beethoven," said Ruth, "but Shostakovich really challenges you into another world again."

A new heaven and a new earth

After these thoughts, we decided that the eternity of the book of Revelation is one of activity, joy, and wonder, inviting us to share in God's new dimension of spacetime, "the new heaven and new earth." Time itself will be redeemed with the whole cosmos. God's eternity has been made known and its mystery opened up in Jesus—the self-involvement of God. "Perhaps the old earth and heavens in their three dimensions will have passed their sell-by time. All the extra seven or so dimensions could then unroll, giving us a thrilling new picture of what happens when this world ends." I reminded them of the prophetic words from the New Testament writer to the Hebrews,[6] echoing the verse from Isaiah:[7]

> In the beginning, O Lord, you laid the
> foundations of the earth,
> And the heavens are the work of your hands.
> They will perish, but you remain;
> they will all wear out like a garment.
> You will roll them up like a robe;
> like a garment they will be changed.
> But you remain the same,
> And your years will never end.

And as my friends inevitably suggested in the Flatland story, "He will lift them all up!"

"Which way to heaven?" challenged Angus. "Do we enter at the Cross, rather than in vague transcendence?" Josh demanded.

"The new heaven could be a universe with more dimensions!" Richard exclaimed. "I think we are beginning to take M-theory seriously!"

We occasionally live in two times: England time, say, and Beijing time ("a world away," I said, checking my "prayer watch" on my right wrist, set eight hours ahead). "It's as if the Creator-God of the universe can see all times at once: certainly earth time, Jupiter time; or time on the other side of the galaxy, 100,000 light-years away." Josh was astounded by his own logic. "He is in more than a one-dimensional chronos. He is the LORD of all time, LORD of all dimensions." God stands outside time as well as within it—"kingdom time," as Jesus might have put it.

Heaven is then God's dimension, beyond and within the world of space, time and matter, which for some is all there is. Now we are invited to see dimensions of our world that are normally hidden, but which Jesus brings to new life. The Spirit is God's presence and power within creation.[8]

Opening doors to the spiritual realm

We seemed ready to say that eternity may be glimpsed from time to time in the natural world, in our consciences, in meeting the Holy Spirit—particularly in healing. "Could it be that 'eternal life' is even more important than physical health?" wondered Josh. Jesus equated it with the kingdom of heaven. Melissa accepted this: "The kingdom of heaven must be here. It starts with the air we breathe, the very room we are in, if we are ready."

"What about the next life?" Angus challenged.

Ruth was paraphrasing the original: "In my Father's house are many dimensions."

"F-theory indeed," thought Melissa and Ruth.

"If we welcome the Spirit, even now God adds a dimension of himself to our spirits." It was this new quality of life, enjoyed with God now and to be enjoyed hereafter, which Josh found such a completely fresh insight. Josh was aware that our bodies, souls, and spirits were intimately joined: "and with the God-given Spirit dimension if we invite him." Melissa agreed. "M really does stand for mystery!"

We may even suggest a hierarchy of dimensions, from the highest, down through angels and archangels, in the biblical model. "Yes," mused Josh, "and the Satan, the Evil One, with his demons or dark spirits in much lower dimensions, already defeated by Jesus in his cross and Resurrection!"

Things were beginning to come together for the Flatlanders. It was Richard and Ruth's turn to challenge us: "This is surely M-theory come down to earth; talk about shadows of the eternal in higher dimensions!" Angus was feeling euphoric: "This is really doing my head in," he laughed.

"If God has prepared these redemptive analogies we heard about, they may be found in everything we have been talking about all the year.

We should expect to find them in today's cosmology and physics." This was to be our last challenge.

Dimensions of the spirit in physics today

Angus remembered the first thing that struck him amidst the awe and wonder of the big bang theory of creation was "if there was a Creation, there must be a Creator." For Ruth the extra embedding dimensions needed to describe curved space gave her room to think of God as enfolding the world with his love and care, embedding us in his Spirit. Josh had found that the idea of the black holes being singularities spoke powerfully to him. "Entry points to other dimensions?" he pondered. "Isn't Jesus the real singularity, an entry point or gateway to the kingdom and to the coming of the Spirit?" "We are all entry points," Ruth almost surprised herself. "All of us are like a singularity, where we can welcome the Spirit when we create space for the Spirit to speak to us" ("or the dark forces," reminded Melissa). For Ruth it was Jesus giving himself on the cross. This is what opened the way, the greatest singularity of all, allowing God to send his Holy Spirit. "The dimensions of the universe were wonderfully expanded at that point!" said Richard, warming to his theme.

As God said to Job, "Even chaos is under my control."

Richard had gone further, remembering the evidence of those fractals, pictures that revealed the wonderful hidden order beneath chaos. "The fractals point straight to Jesus," he saw clearly. "The amazing hidden order within the apparent chaos is what John called the *Logos* (and the Chinese call the Tao) and the *Logos* made flesh was Jesus, of course."[9] Richard, looking further, saw Jesus as the real Attractor. "Perhaps a better analogy is the Hidden Attractor, as he is not always recognized."

It is even said that evolution is chaos with feedback and a hidden order. Adam and Eve represent the first *Homo sapiens* to become aware of the spiritual dimensions beyond the physical.

But we all knew: it was Flatland that changed us. Particularly as there was no religious language and that they themselves unraveled what was to happen. As Angus remarked, "It seemed to take on a life of its own." Ruth reminded us that the Sphere didn't just send down a circle. "He came himself, and took the risk of openness and vulnerability, and that the section of the Sphere seen as the circle was of course Jesus for us."

"He had to put off the glory of many dimensions and become a Flatlander."

"Certainly the new picture from today's physics of the world, as only a three-dimensional shadow of a deeper reality of many dimensions, sort of stabs you alive!" Richard was ready for the final frontier of quantum theory. "This deeper reality is observer-centered and the observer is Jesus," Josh confirmed, "and the pilot-wave or guiding wave in higher dimensions is the Spirit." Einstein's "spooky action at a distance" of particles, once connected, comes to life in extra dimensions; "extra dimensions of prayer" was Ruth's analogy. We cannot now be separated from the wider dimensions of God's love. The extra or higher dimensions certainly illuminate what M-theory describes in the new duality.

"Wow!" exclaimed Richard, "I can see how the idea of a particle being at the same time also a probability wave in more than three dimensions is seen as the same underlying concept in duality. Wave and particle, like the different string theories, describe exactly the same physics, and are really aspects of the same underlying reality."

Melissa was ecstatic: "Now we can see how this illuminates the creedal belief in Jesus as both divine and human.[10] What a wonderful revelation from today's physics—now that's a redemptive analogy!"

"It all fits together!"

A deeper reality for both science and belief. A new unification rather than the old conflict. The cost of ignoring this is a distant God, a loss of immediacy of the presence of the Holy Spirit and of the encounter with Jesus.

Redemptive analogies abound in waterfalls and blast furnaces, harbors, and mountaintops, a vision over Beijing, "or even the Harry Potter books," added Melissa. "We have to be real to ourselves, our spiritual experiences, our mind and reason."

The Spirit reveals himself through the created order, in cosmology and physics, as well as through our own experience of his presence. Heaven is "God's dimension of present reality"—a dimension, normally kept secret, a deposit, a gift for our encountering the Spirit, if we are willing. Perhaps chaos and disorder are like suffering. Inner healing then leads to a new wholeness and integration, unification between the Lord and ourselves if we become a new creation in Jesus. Through him we can "know the mind of God."

I predict that taking M-theory seriously is how scientists will investigate all areas at the sharp end of science today. Science and faith will be seen as mutually compatible insights within a multidimensional universe, the new worldview of contemporary science.

Shalom[11] between science and belief!

Indeed, between physics and the Holy Spirit!

Glossary

Alpha particles The nuclei of helium atoms emitted by radioactive materials.

Anthropic principle An explanation for why the universe has the properties we observe is that if it were different, life would not form, and therefore we would not be here to observe it. (See chapter 6)

Antimatter Matter that has the same gravitational properties as ordinary matter, produced only in particle accelerators (and in the big bang). It is a mirror-image kind of matter with opposite electric or nuclear charges. When a particle meets with its antiparticle, they annihilate, leaving only energy.

Atom Fundamental building block of matter, made up of a tiny nucleus (consisting of protons and neutrons) surrounded by a swarm of orbiting electrons.

Attractors In a complex system, these represent the states to which the system eventually settles, depending on the properties of the system.

Big bang Currently widely accepted theory that the expanding universe began about 14,000 million years ago from a state of enormous energy, density, and compressed into a tiny point, a singularity. (See chapter 1)

Big crunch One hypothesized future for the universe in which the current expansion stops, reverses, and results in all matter and space collapsing together to a singularity, a reversal of the big bang.

Black hole An object that has such an enormous gravitational field that not even light can escape from it.

Brane Any of the extended objects that arise in string theory. A one-brane is a string, a two-brane is a membrane, a three-brane has three extended dimensions, etc. In general, a p-brane has "p" spatial dimensions.

Compactified dimensions Highly curved dimensions of string theory, rolled up into a very tiny space.

Cosmic microwave background radiation The radiation from the glowing of the hot early universe of the big bang, now so greatly thinned and cooled (3°K) that it appears not as light but as microwaves (radio waves with a wavelength of a few centimeters).

Cosmological constant A mathematical device used by Einstein to give spacetime an inbuilt ability to expand.

Cosmology The study of the universe as a whole.

Curled-up dimension A spatial dimension that is crumpled, wrapped, or curled up into a very tiny size, so cannot be directly observed.

Curved space The deviation of space or of spacetime from the flat form described by Euclidean geometry.

Dark matter The universe is permeated with matter that is invisible to the astronomer's telescope. No one is certain of its identity, although there are several candidates for the dark matter required to fit the evidence that it is needed.

Dimension An independent direction in space or spacetime. The familiar space around us has three dimensions (up, down, and sideways/back and forth). The familiar spacetime has the extra time or past-future axis. Superstring theory requires the universe to have additional spatial dimensions.

Duality A deep and surprising symmetry where two or more theories appear to be completely different, yet are identical at a deeper level, and actually give rise to the same physical results.

Electro-encephalograph An instrument recording the patterns of electrical impulses in the brain, used for diagnosis (also EEG).

Electromagnetic force One of the four fundamental forces, produced by the interaction of charged particles (chiefly electrons) with magnetic fields.

Electron A particle with a negative unit charge that orbits the nucleus of an atom.

Encephalisation The process of increasing brain size in early primitive mankind.

Eukaryotic cell The nucleus of the cell contains the genetic material DNA, and is surrounded by mitochondria—threadlike bodies that produce energy.

F-theory (father theory) All-encompassing superstring theory in twelve dimensions, ten space and two time.

Fermion A particle, or a pattern of vibrating strings—the familiar protons and neutrons and their constituent quarks.

Fractals A word coined by Mandelbrot in 1975, geometrical shapes that, contrary to Euclid, are irregular all over. In fact, they have the same degree of irregularity on all scales. Any subsystem of a fractal system is equivalent to the whole: it exhibits "self-similarity."

Fundamental forces Electromagnetism, gravity, the strong force binding the nucleus of the atom, and the weak force of radioactivity.

General relativity Einstein's theory based on the idea that the laws of science should be the same for all observers, no matter how they are moving. It explains the force of gravity in terms of the curvature of a four-dimensional spacetime.

Grand unified theory A theory that unifies the electromagnetic, strong, and weak forces.

Inflation, inflationary cosmology Modification to the earliest moments of the standard big bang cosmology in which the universe undergoes a brief burst of enormous expansion.

Leptons The world as we know it is made of leptons, the class including electrons and quarks.

Light-year The distance traveled by light in one year.

M-theory Theory emerging from the second superstring revolution, which unites the five previous superstring theories within a single overarching framework. It appears to involve eleven spacetime dimensions.

Neuroscience The scientific analysis of the way the brain works by examining the processes involving the neurons and nerve cell attachments.

Neutrino An extremely light (possibly massless) elementary matter particle, which is affected only by the weak force.

Neutron An uncharged particle, typically found in the nucleus of an atom (along with roughly the same number of protons).

Nucleus The core of the atom, consisting of protons and neutrons, held together by the strong force.

Nuclear fusion The process in which two nuclei collide and coalesce to form a single heavier nucleus, often emitting a great deal of energy.

Particle accelerator Machine for boosting particles to nearly the speed of light and slamming them together in order to probe the structure of matter.

Phase space A mathematical representation of more than three dimensions into a lower-dimensional picture—a projection of "mappings," usually on to three dimensions. A phase portrait is like a cross section or slice (as an aerial view of the ground is a way of reducing three dimensions to two).

Proton Positively charged particle, typically found in the nucleus of an atom, and consisting of three quarks.

Quantum mechanics The framework of laws governing the universe whose unfamiliar features such as uncertainty, quantum fluctuations, and wave/particle duality become most apparent at the microscopic scales of atoms and subnuclear particles.

Quark A charged elementary particle acted upon by the strong force, stuck together by interaction between "gluons." Protons and neutrons are each composed of three quarks; these exist in six varieties (up, down, charm, strange, top, bottom) and three "colors" (red, green, and blue).

Quasars Strange astronomical objects (bright and strongly red-shifted) believed to be very distant, very intense sources, perhaps deriving their energy from the presence within them of a very massive black hole.

Red shift The reddening of light from a star that is moving away from us, due to the Doppler effect.

Redemptive analogy The centrality of Jesus seen, not as an alien concept, but as the fulfillment of the best in each local custom or culture. God has also laid down an "inside track" for scientists even within modern physics.

Singularity A point in spacetime at which the spacetime curvature becomes infinite. Such naked singularities might provide entrances to tunnels ("wormholes") through the fabric of spacetime.

Strange attractors A class of attractors known as "chaotic" or strange attractors, existing in multidimensional phase space. They are both attractors and exhibit sensitive dependence on initial conditions. They reveal complex behavior that seems random but actually has some hidden order.

String theory Unified theory of the universe postulating that the fundamental ingredients of nature are not zero-dimensional point particles but tiny one-dimensional filaments called strings. String theory harmoniously unites quantum mechanics and general relativity, the previously known laws of the small and the large, which are otherwise incompatible.

Supernova The nuclear fusion reactions in some large stars can run away to produce a prodigious explosion—as much as is usually produced in a whole galaxy of, perhaps, a thousand million stars like our sun. The outer layers, including heavy metals produced in the interior, are blown off, leaving behind a very compact residue, often seen as a pulsar.

Superstring theory Often shortened to "string theory," and which incorporates supersymmetry.

Torus The two-dimensional surface of a doughnut.

White hole Just as matter can collapse into a black hole singularity, so a singularity might pour matter outward into our universe in the cosmic gusher of a white hole.

Wormhole A tubelike region of space connecting one region of the universe to another.

Notes

Chapter 1: The Universe and Beyond: How Did It All Begin?

1. John D. Barrow, *Theories of Everything* (Oxford: Oxford University Press, 1991), 80.
2. Alan Guth, *The Inflationary Universe* (1981; repr., London: Jonathan Cape, 1997). Guth is a physicist associated with the Massachusetts Institute of Technology.
3. Steven Weinberg gives a beautiful account of this in his book *The First Three Minutes* (London: Andre Deutsch, 1977), 12.
4. Stanley L. Jaki, *God the Cosmologist* (Edinburgh: Scottish Academic Press, 1989).
5. Martin Rees, "Ripples from the Edge of Time," *The Guardian,* April 23, 1992, 19; George Smoot was jointly awarded the Nobel Prize for Physics on October 13, 2006, referenced in Alok Jha, "Big Bang Theory Physicists Share Nobel Prizes," *The Guardian,* Oct. 14, 2006.
6. Bernard Lovell, "The View of Life from the Edge of the Universe," *The Observer,* August 23, 1992, report on a meeting of the American Physical Society held on April 23, 1992.
7. Douglas Clowe in Zeeya Merali, "Dark Matter Gets Its Own Dark Force," *New Scientist,* January 20, 2007, 12. This could be confirmed when the LSST sees first light in 2013 (Large Synoptic Survey Telescope), in collaboration with Google. This would confirm the role of dark matter as the scaffolding round which visible stars and galaxies congregate.
8. Douglas Clowe and Dennis Zaritsky, "Shot in the Dark," *Physics World,* February 2007, 26–28.
9. Paul Davies, *The Last Three Minutes* (London: Weidenfield and Nicholson, 1994), 89.
10. Personal communication, September 15, 1980.

Chapter 2: Mysteries, Models, and Quantum Theory

1. Niels Bohr, *Atomic Physics and Human Knowledge* (New York: John Wiley, 1958).
2. Roger Penrose, *The Emperor's New Mind* (London: Vintage, 1990), 385.

3. John A. Wheeler, "John Wheeler" in P. C. W. Davies and J. R. Brown, eds., *The Ghost in the Atom* (Cambridge: Cambridge University Press, 1986), 60.
4. Albert Einstein in private letters to his friend James Franck, in John Stachel, ed., *The Collected Papers of Albert Einstein* 7, Einstein Papers Project, Boston University (Princeton, N.J.: Princeton University Press, 2001).
5. Murray Gell-Mann's speech when accepting the 1969 Nobel Prize for physics for his classification of the elementary particles known as quarks. Murray Gell-Mann, *The Quark and the Jaguar* (New York: Viking, 1997), introduction.
6. Cited in J. M. Cohen and M. J. Cohen, *Penquin Dictionary of Quotations* (New York: Penguin, 1960).
7. Richard Feynman, *The Character of Physical Law* (Cambridge, Mass.: MIT Press, 1995), 1. Feynman's views are succinctly presented in his series of television lectures at Cornell University in 1964 and broadcast on BBC2 in 1965. These lectures have been republished as *Six Not-so-easy Pieces* (London: Allen Lane, 1999).
8. Penrose, *Emperor's New Mind*, 385, 523.
9. Fritjof Capra, *The Tao of Physics* (London: Fontana/Collins, 1976); Gary Zukav, *The Dancing Wu Li Masters* (London: Flamingo/Fontana, 1980).
10. Bruce Rosenblum and Fred Kuttner, *Quantum Enigma: Physics Encounters Consciousness* (London: Duckworth, 2006).

Chapter 3: Quarks, Supersrings, and M-branes: A Theory of Everything for the Twenty-first Century?

1. Stephen Hawking, *A Brief History of Time* (London: Bantam, 1988), 175.
2. Ed Witten, "The Pied Piper of Superstrings," *Scientific American*, November, 1991, 18. See also BBC Radio 3 program, "Desperately Seeking Superstrings," broadcast on February 14, 1987.
3. Personal conversation, August 10, 1985.
4. Letters from Einstein to Kaluza dated April 28, 1919; May 29, 1919; and October 14, 1921; in John Stachel, ed., *The Collected Papers of Albert Einstein* 7, 1918–1921 (Princeton, N.J.: Princeton University Press, 2001). I have copies of these letters in my possession, but the originals are still in the possession of the Kaluza family.
5. For further details, see Eric W. Middleton, "Higher Dimensional Theories in Physics Following the Kaluza Model of Unification" (master's thesis, University of Durham, n.d.).
6. Personal communication, June 10, 1982.
7. Personal communication, January 10, 1986.
8. Brian Greene, *The Elegant Universe* (New York: Jonathan Cape, 1999), 317.
9. Lisa Randall, "Why I Believe in Higher Dimensions!" *Daily Telegraph*, June 1, 2005, 14; and Lisa Randall, interview by Anna Fazackerley, "I'm Fairly Confident New Dimensions Are Out There," and "Trust me, I *am* a

Physicist," *Times Higher Educational Supplement,* June 3, 2005. These articles were in advance of her speaking at the Cheltenham Science Festival on Higher Dimensions, June 11, 2005. See also her book *Warped Passages: Unraveling the Mysteries of the Universe's Hidden Dimensions* (New York: Allen Lane/Penguin, 2005).

10. John A. Wheeler and Kip Thorne, *A Journey into Gravity and Spacetime* (San Francisco: W. A. Freeman, 1990), 149, 209. See also, Stephen Hawking, *The Universe in a Nutshell* (London: Bantam, 2001), 135–37.
11. Michio Kaku, "Will We Ever Have a Theory of Everything?" *New Scientist,* November 18, 2006, 65.
12. Sheldon Glashow and Paul Ginsparg, "Desperately Seeking Superstrings," *Physics Today*, May 1986, 7.

Chapter 4: What Is Reality?

1. Max Planck, *The Universe in the Light of Modern Physics* (London: Allen and Unwin, 1931), 8.
2. For the full allegory, see Book 7 of Plato's *Republic,* translated by H. D. P. Lee (New York: Penguin, 1955).
3. William James, *The Varieties of Religious Experience,* The Gifford Lectures, Edinburgh, 1901–2 (London: Fontana/Collins, 1960), 464.
4. William James, *Essays in Pragmatism* (1920; New York: Free Press, 1970), chapter 1.

Chapter 5: The Story of Flatland

1. A. Square (alias E. A. Abbott) wrote the nineteenth-century novel *Flatland: A Romance of Many Dimensions* (1884, U.K.: Penguin Books, 1986). This book has been published by numerous publishers as well as translated into many languages. Banish Hoffmann wrote the introduction for the 1986 Penguin edition.
2. The real enjoyment of *Flatland* is that this metaphor rapidly took on a life of its own. All I had to do was ask "what happens next," and the story unfolded. It gave us language for talking openly about issues of faith and doubt.

Chapter 6: The Anthropic Principle

1. Quoted in John Horgan "The Return of the Maverick," *Scientific American,* March 1995, 25.
2. This principle was first clearly stated by Brandon Carter in "Large Number of Coincidences and the Anthropic Prinicple in Cosmology" in *Confrontation of Cosmological Theories with Observational Data,* ed. M. Longair (Guildford, U.K.: Springer London, 1974).

3. Roger Penrose, *The Emperor's New Mind* (London: Vintage, 1990), 445.
4. G. T. Whitrow, "Why Physical Space Has Three Dimensions?" *British Journal for the Philosophy of Science* 6 (1955): 345–46.
5. Sir Martin Rees and John Gribbin, "Cosmic Coincidences," *New Scientist* January 13, 1990, 54. Helium is important because, as Fred Hoyle showed, higher elements like carbon can only be formed by a complicated process involving the combining of helium nuclei in two stages.
6. For more information on this topic, see Peter D. Ward and Donald Brownlee, *Rare Earth* (New York: Springer, 2000) and Guillermo Gonzalez, Donald Brownlee, and Peter D. Ward, "Refuges for Life in a Hostile Universe," *Scientific American*, October 2001, 52–59.
7. Stephen Hawking, *A Brief History of Time* (London: Bantam, 1988), 174–75.
8. For more information about his argument, see Brandon Carter, "Large Number Coincidences and the Anthropic Principle," in *Confrontation of Cosmological Theories with Observation*, ed. M. S. Longair (Dordrecht: Reidel, 1974), 291; and Brandon Carter, "The Anthropic Principle and Its Implications for Biological Evolution," in *The Constants of Physics*, ed. W. McCrae and M. J. Rees (London: The Royal Society, 1983), 137.
9. Hermann Bondi, Nobel Prize winner in mathematics, was also president of the British Humanist Association. For this comment, see John Wheeler in P. C. W. Davies and J. R. Brown, eds., *The Ghost in the Atom* (Cambridge: Cambridge University Press, 1986), 60.
10. Fred Hoyle, quoted in P. C. W. Davies, *The Accidental Universe* (Cambridge: Cambridge University Press, 1982), 118.
11. Horgan, "The Return of the Maverick," 24.
12. Paul Davies, *The Edge of Infinity* (London: Dent, 1981), 161, 171.
13. Rees and Gribbin, "Cosmic Coincidences," 54.

Chapter 7: Evolution: Where Do We Fit In?

1. R. J. Berry (professor of genetics at University College, London), *God and Evolution* (London: Hodder and Stoughton, 1988), 163.
2. Bernard Wood, "A Precious Little Bundle," *Nature*, September 21, 2006.
3. For further discussion of this, see Berry, *God and Evolution*, 72. See also J. S. Jones and S. Rouhani, "How Small Was the Bottleneck?" *Nature* 319 (1986): 449.
4. Celia Deane-Drummond, *Wonder and Wisdom* (Philadelphia: Templeton Foundation Press, 2006), 80.
5. Keith Ward, *God, Faith and the New Millenium* (Oxford: Oneworld, 1998), 110, 111.
6. Charles Darwin, *The Origin of Species* (1857; repr., London: Everyman Library, Dent, 1928), 463.
7. Deane-Drummond, *Wonder and Wisdom*, 69.

8. Simon Conway Morris, *Life's Solutions: Inevitable Humans in a Lonely Universe* (Cambridge: Cambridge University Press), 330, 304.
9. Richard Dawkins, *The Selfish Gene* (Oxford: Oxford University Press, 1989); and *The Blind Watchmaker* (New York: Penguin, 1991).
10. Deane-Drummond, *Wonder and Wisdom*, 28.
11. Stephen J. Gould, *Wonderful Life* (London: Hutchinson Random, 1990), 560.
12. Berry, *God and Evolution*, 97, 98.
13. Roger Penrose, *The Emperor's New Mind* (London: Vintage, 1990), 537–38.
14. Berry, *God and Evolution*, 163.
15. Richard J. Clifford and Roland E. Murphy, "Genesis," in *The New Jerome Bible Commentary* (London: Geoffrey Chapman, 1989), 11.

Chapter 8: Consciousness and What Comes After

1. Susan Greenfield, *The Human Brain: A Guided Tour* (London: Weidenfeld and Nicholson, 1997; reissued, London: Phoenix, 2000), 192.
2. Law of Complexity-Consciousness in Teilhard de Chardin, *Man's Place in Nature* (London: Collins/Fontana, 1971), 98.
3. Greenfield, *Human Brain*, 139.
4. Ibid., 192.
5. Euan Squires, *To Acknowledge the Wonder* (Bristol, U.K.: Adam Hilger, 1985).
6. Roger Penrose, *The Emperor's New Mind* (London: Vintage, 1990), 523.
7. Sir John Eccles (d. 1997) was a distinguished Australian neurophysiologist who shared the 1963 Nobel Prize for physiology. He made fundamental discoveries concerning the interaction of neurons (nerve cells) in the transmission of a nerve impulse. For more information about his thinking, see J. C. Eccles, *The Neurophysiological Basis of the Mind* (Oxford: Clarendon Press, 1953). He himself recommended (private correspondence, May 21, 1980) *The Human Psyche* (London: Springer, 1980), lecture 10.
8. J. R. Smythies, *Analysis of Perception* (London: Routledge and Kegan Paul, 1956) is a discussion of physical and perceptual space. See also Smythies, "Analysis of Projection," *British Journal for the Philosophy of Science* 5 (1954): 120.
9. Charles D. Broad of Cambridge University was a contemporary of Russell and Wittgenstein. For his ideas on this topic, see his book *Scientific Thought* (London: Routledge and Kegan Paul, 1927), 543. His ideas are also quoted by Sir Russell Brain in *The Nature of Experience* (Durham: University of Durham, 1959), 35.
10. John Polkinghorne, *Science and Creation* (London: SPCK, 1988), 72.
11. Luke 17:21.
12. Tom Wright (Canon Theologian of Westminster Abbey), *Luke for Everyone* (London: SPCK, 2001), 210.

13. Matthew 25:31–46; Luke 10:30–37.
14. John Habgood, former archbishop of York, was also president of the Science and Religion Forum. For him, the sacramental principle applied to the whole of nature. Ultimate meaning is found as men find a purpose for their lives in Jesus. See John Habgood, *Religion and Science* (London: Hodder and Stoughton, 1972).
15. Michael Ramsey, former archbishop of Canterbury, in a sermon delivered in St. John's College chapel in May 1983, and *Be Still and Know* (London: Fount Paperbacks, 1982), 57.
16. Revelation 20:14.
17. John Stott writes: "The biblical evidence for the ultimate annihilation of those who reject Jesus is stronger than is commonly thought." See David Edwards and John Stott, *Essentials: A Liberal-Evangelical Dialogue* (London: Hodder and Stoughton, 1988), 319–20.
18. David Jenkins (previously bishop of Durham), *Free to Believe* (London: BBC Books, 1991), 94.
19. George Carey, *Canterbury Letters to the Future* (Eastborne, East Sussex, U.K.: Kingsway Publications, 1998), 220, 221. "To choose hell is to choose the rubbish heap, the place of thrown away things, with all the pain associated with waste." A second picture is of "eternal separation from God our Father," "an image terrible in its desolation of loneliness and finality." He warns, however, that "it is important not to confuse the vivid descriptions that the ideas come dressed in, with the realities they express."
20. Matthew 18:21–35.
21. Luke 24:13–35.
22. John 20:19.
23. 1 Corinthians 15:35–44.
24. James D. G. Dunn (Lightfoot Professor of Divinity at the University of Durham), *Jesus and the Spirit* (London: SCM Press, 1975), 308.
25. C. S. Lewis, *The Great Divorce* (London: Geoffrey Bles, 1946; London: Fount Paperbacks, 1997). He also wrote that hell cannot compare with "the joy that is felt by the least in Heaven . . . For a damned soul is nearly nothing: it is shrunk, shut up in itself" (104).
26. John 20:28.
27. Dunn, *Jesus and the Spirit*, 132.
28. A good place to start your enquiry might be with James Dunn, *The Evidence for Jesus* (London: SCM Press, 1985), or with David Runcorn, *Rumours of Life* (London: Darton, Longman and Todd, 1996).
29. Rowan Williams, *Resurrection* (London: Darton, Longman and Todd, 1982), 70.
30. Dunn, *Jesus and the Spirit*, 114, 129.

Chapter 9: Miracles and Missions

1. Tom Wright, *Luke for Everyone* (London: SPCK, 2001), 311.
2. Exodus 14:9–31. See also Colin Humphreys, *The Miracle of Exodus* (San Francisco: HarperSanFrancisco, 2003).
3. Graham Dow, now bishop of Carlisle, has written a book on this, called *Explaining Deliverance* (Chichester, U.K.: Sovereign World, 1991).
4. Wright, *Luke for Everyone*, 311.
5. John 6:35.
6. John 2:1–11.
7. William James, *The Varieties of Religious Experience*, The Gifford Lectures, Edinburgh, 1901–2 (London: Fontana/Collins, 1960), 464.
8. For an excellent overview, see Ninian Smart's book, *The Religious Experience of Mankind* (London: Fontana, 1971).
9. James, *The Varieties of Religious Experience*, 484.
10. Comment made at the interfaith conference on spirituality for the twenty-first century, "Social and Spiritual Regeneration," held in Newcastle upon Tyne, June 1998.
11. Ambrose Griffiths, Roman Catholic bishop of Newcastle. Comment made at the interfaith conference on spirituality for the twenty-first century, "Social and Spiritual Regeneration," held in Newcastle upon Tyne, June 1998.
12. John V. Taylor (bishop of Winchester), *The Go-Between God* (London: SCM Press, 1972), 190.
13. Mahesh Chavda is a converted Hindu with an international healing ministry based on fasting and prayer in the power of the Holy Spirit. See *Watch of the Lord* (Lake Mary, Fla.:Creation House, 1999).
14. Don Richardson, *Peace Child* (Ventura, Calif.: Regal Books, 1976), 234.
15. Don Richardson, *Eternity in their Hearts* (Ventura, Calif.: Regal, 1981), 60.
16. Taylor, *Go-Between God*, 190.
17. David Jenkins, *Free To Believe* (London: BBC Books, 1991), 59.
18. Taylor, *Go-Between God*, 196.

Chapter 10: Chaos and the Hidden Order

1. Stephen Hawking, *A Brief History of Time* (London: Bantam, 1988), 122.
2. Benoit B. Mandelbrot, *The Fractal Geometry of Nature* (London: W. H. Freeman, 1982), summarized in Benoit Mandelbrot, "Fractals: A Geometry of Nature," *New Scientist*, September 15, 1990, 39, reprinted in Nina Hall, ed., *The New Scientist Guide to Chaos* (New York: Penguin Books, 1991), 125.
3. Walter J. Freeman, "The Physiology of Perception," *Scientific American*, February 1991, 78.
4. Reza Tavakol, "Is General Relativity Fragile?" (Third International Conference on the History and Philosophy of General Relativity, University of Pittsburgh, 1991).

5. For more information, see M. Mitchell Wardrop, *Complexity* (New York: Touchstone/Simon and Schuster, 1992).
6. Mandelbrot, "Fractals," 38.
7. Paul Davies, *Superforce* (London: Heinemann, 1984), 243.
8. John Barrow, *The World Within the World* (Oxford: Oxford University Press, 1988), 361.
9. Colossians 1:15.
10. Deuteronomy 18:15.
11. Isaiah 53:1–12.
12. Ephesians 1:10.
13. John 1:1.
14. John 1:14.
15. John 14:9, RSV.

Chapter 11: But Why? The Problem of Evil

1. Jack Deere, *Surprised by the Power of the Spirit* (Eastbourne, U.K., Kingsway Publications, 1993), 93.
2. Stanley Milgram's classic experiments, described in his book *Obedience to Authority* (London: Collins, 1963), have been repeated in Australia, South Africa, and several European countries. In one study conducted in Germany, over 85 percent of the subjects administered a lethal electric shock to the learner (an actor). In 1971, a psychologist, Philip Zimbardo, divided eighteen student volunteers into "guards" and "prisoners." Within six days the guards' behavior had become so brutal that the experiment was stopped; see "The Experiment," *The Guardian*, October 16, 2001, 2–3.
3. For the full story, read Genesis 3.
4. John Polkinghorne, *The Way The World Is* (London: SPCK, 1983), 103.
5. Genesis 3:8.
6. "Jung calls the other side of ourselves, which is to be found in the personal unconscious, the shadow . . . There is no sun without the shadow." See Frieda Fordham, *An Introduction to Jung's Psychology* (New York: Penguin Books, 1953), 50.
7. Lt. General Roméo Dallaire speaking at the National Prayer Breakfast in Ottawa, May 10, 2001.
8. Sir Norman Anderson was professor of law at the Institute of Advanced Legal Studies, London.
9. Joyce Huggett, *Listening to Others* (London: Hodder and Stoughton, 1988). The chapters that spoke to me most were chapter 8, "Listening to Past Pain," and chapter 9, "Making Peace with the Past."
10. Paul Y. Cho, *Prayer: Key to Revival* (Milton Keynes, U.K.: Authentic Media, 1984), 43, 44.

11. Walter Wink, quoting Dr. Raimundo Valenzuela of Santiago, Chile, who was present when this was said, in *Unmasking the Powers* (Philadelphia: Fortress Press, 1986), 54.
12. Peter Zimbardo, *The Lucifer Effect: How Good People Turn to Evil* (London: Rider, 2007).

Chapter 12: The View from Here

1. Colossians 2:17, NIV.
2. John Drane, Ross Clifford, and Philip Johnson, *Beyond Prediction* (Oxford: Lion Publishing, 2001).
3. Ephesians 6:10–18, NIV. I have paraphrased some of the verse, placing it in parenthesis.
4. For a vivid and humorous account of some of the techniques he uses, see C. S. Lewis, *The Screwtape Letters* (London: Fount Paperbacks, 1977).

Chapter 13: M-theory in Eleven Dimensions: The Spiritual Universe

1. Brian Greene, *The Elegant Universe* (New York: Jonathan Cape 1999).
2. Stephen Hawking, *The Universe in a Nutshell* (London: Bantam, 2001), 54, 55.
3. Martin Rees, *The Observer*, magazine supplement, December 13, 2000.
4. Gregory E. Ganssle, ed., *God and Time* (Downers Grove, Ill.: InterVarsity, 2001), 14, 15.
5. Jeremy S. Begbie, *Theology, Music and Time* (Cambridge: Cambridge University Press, 2000), 129.
6. Hebrews 1:10–12. See also N. T. Wright, *New Heavens, New Earth* (Nottingham, U.K.: Grove Books, 1999).
7. Isaiah 51:6.
8. Tom Wright, *Mark for Everyone* (London: SPCK, 2001), 232.
9. John 1:1, 14.
10. Our new insight through M-theory and eleven or twelve dimensions shone great light on the truth of the "Chalcedonian definition" concerning the two natures of Christ, which underpins our whole exploration. This is held by many to be one of the peaks of intellectual achievement in classical antiquity. The Council of Chalcedon correctly ensured that Jesus was acknowledged as both divine and human, although at that time further exposition was not possible.
11. *Shalom*—basic meaning is "wholeness, completeness," but implies much more, e.g., "peace offering" as part of restoring a broken relationship.

Index